GW01184418

AN INTRODUCTORY GUIDE TO PUMPS AND PUMPING SYSTEMS

An Introductory Guide to Pumps and Pumping Systems

R. K. TURTON

Department of Mechanical Engineering
Loughborough University of Technology

Series Editor
Roger C. Baker

Mechanical Engineering Publications Limited, London

First published 1993

ISBN 0 85298 882 6

A CIP catalogue record for this book is available from the British Library.

Typeset by MFK Typesetting Ltd, Hitchin
Printed and bound by Page Bros, Norwich

SERIES EDITOR'S FOREWORD

As an engineer I have often felt the need for introductory guides to aspects of engineering outside my own area of knowledge. MEP welcomed the concept of an introductory series to follow on from my own book on flow measurement. We hope that the series will provide engineers with an easily accessible set of books on common and not-so-common areas of engineering. Each author will bring a different style to his subject, but some valued features of the original volume, such as conciseness and the emphasis of certain sections by shading, have been retained. The initial volumes are biased towards fluids, but we hope to broaden the scope in later volumes.

The series is designed to be suitable for practising engineers and technicians in industry, for design engineers and those responsible for specifying plant, for engineering consultants who may need to set their specialist knowledge in a wider engineering context, and for teachers, researchers, and students. Each book will give a clear introductory explanation of the technology to allow the reader to assess commercial literature, to follow up more advanced technical books and to have more confidence in dealing with those who claim an expertise in the subject.

Keith Turton, the author of this second volume in the series, is widely known and respected as an authority on pumps. He combines a theoretical knowledge with an industrial and commercial awareness, as a result of his wide experience in consulting for industry and in directing a highly successful short course which has run for about 20 years and has benefited over 1,500 participants from industry.

I would particularly like to acknowledge the encouragement of Mick Spencer, the Managing Editor. We both hope that the series will find a welcome with engineers and we shall value reactions and any suggestions for further volumes in the series.

Roger C. Baker
St Albans

Other titles in this series:

An Introductory Guide to Flow Measurement – Roger C. Baker

CONTENTS

AUTHOR'S PREFACE

An engineer choosing a pump for a given application must define the duty (or duties), determine the type of pump to be used, the materials of construction, and how it should be driven. Of course, in many cases, a standard pump is all that is required, but the buyer still needs to be sure that the pump is 'fit for duty'. To do this the engineer needs to know the basic criteria involved in making this judgement.

This book is thus designed to give an understanding of the types of pumps available, their operating principles, and the way they interact, with the necessary background to allow the engineer to assess information from manufacturers and make a contribution to technical discussions.

The engineer using a fluid system always has to decide how flow is to be provided. If he can place the system on the side of a hill and thus allow gravity to cause flow he is fortunate. In most cases, however, a pump has to be provided to overcome the system resistance to the flow, and it is then necessary to make an accurate estimate of the flow losses, and then to decide what type and size of pump are required. In the chapters which follow, the main types of pump in use in the process industries are described, the principles which underlie their operation are discussed, and the way they interact with the systems they would supply is outlined. The effect of cavitation is described, and the last chapter describes the underlying principles which guide the selection of the correct machine for the duty.

The approach used assumes some basic knowledge of fluid mechanics, but deals with all the essential equations in context. The experience of the author and of colleagues in covering this material in courses for process and plant engineers given at Cranfield Institute of Technology has been distilled into this volume, and it is hoped the information, while concise, is complete enough for the engineer.

If you, the reader, feel that matters of importance have not been discussed clearly or not included, the author will be very pleased to hear from you.

Keith Turton
Loughborough

NOMENCLATURE

Symbol	Description	Unit
A	Area	m^2
b	Passage height	m
c	Acoustic velocity	m/s
c	Blade chord	m
C_L	Lift coefficient	
C_D	Drag coefficient	
D	Diameter of machine element	m
D	Drag force	N
d	Pipe diameter	m
f	Darcy friction factor	
F_A	Force acting in axial direction on blade	N
F_T	Force acting in the tangential direction on blade	N
g	Acceleration due to gravity	m/s^2
gH	Specific energy	J/kg
H	Head	m of liquid
k	Surface roughness	mm
K	Loss coefficient	
k_S	Dimensionless specific speed	
k_{SS}	Dimensionless Suction Specific Speed	
L	Length of pipe	m
L	Lift force	N
$\dot{m}$	Mass flow rate	kg/s
N	Rotational speed	r/min
$NPSE_A$	Net Positive Suction Energy Available	J/kg
$NPSE_R$	Net Positive Suction Energy Required	J/kg
$NPSH_A$	Net Positive Suction Head Available	m of liquid
$NPSH_R$	Net Positive Suction Head Required	m of liquid

Symbol	Description	Unit
N_S	Specific Speed	
p	Pressure	N/m^2
P	Power	W
Q	Volumetric Flow Rate	m
Re	Reynolds number	
r	Bend radius	m
s	Blade spacing	m
U	Peripheral velocity ($D/2$)	m/s
V	Absolute velocity	
V_A	Axial component of absolute velocity	m/s
V_R	Radial component of absolute velocity	m/s
V_U	Tangential component of absolute velocity	m/s
W	Relative velocity	m/s
α	Angle of attack	degrees
β	Angle made by relative velocity vector	degrees
η	Efficiency	
η_O	Overall efficiency	
η_V	Volumetric efficiency	
μ	Absolute viscosity	kg/m s
υ	Kinematic viscosity	m^2/s
ρ	Density	kg/m^3
σ	Thoma's cavitation parameter	
φ	Specific energy coefficient (gH/U^2)	
ω	Angular velocity	rad/s

Subscripts 1, 2, etc., indicate the point of reference. Other symbols of particular significance are defined in the text where they occur.

CHAPTER 1

Introduction to Pump Types and their Layout

1.1 INTRODUCTION

Pumps are used to impart energy to systems that convey a wide range of fluids, which can be listed in three groups of duties to illustrate why a large range of pump designs are available.

Table 1. Pumping duties

Fluid Transference Duties
Blood and medical fluids
Chemical and petro-chemicals
Pharmaceuticals (toothpaste, medicine, etc.)
Water supply, sewage, and effluent
Fire fighting
Boiler feed supply
Heating and refrigeration
Liquid food and drink
Suspended solids (concrete pumping, etc.)
Suspended foods (potatoes, beans, etc.)
Power Transference Duties
Hydraulics (brakes, controls, etc.)
Processing Duties
Water jetting (bark removal, descaling, concrete cutting, etc.)
Site de-watering

There are two main groups of pumps which will be covered in this book, plus 'special effect' machines.

Table 2. The main pump families

Rotodynamic Pumps	*Positive Displacement Pumps*
Centrifugal	Reciprocating
Axial	Rotary
Mixed flow	
Special Effect Pumps	
Jet pumps	
Electromagnetic	
Regenerative	

As Table 1 indicates, the fluids pumped vary considerably in viscosity and density, and range from clean and single phase, through liquids close to boiling point, as in condensate systems, to multi-phase, as in oil-field pumping plant.

In this chapter the design, layout, and performance envelope of the commonly used pumps will be discussed. The chapter ends with a statement of the typical flow and pressure ranges that may be delivered by the pumps which have been described.

The operating principles are discussed in later chapters, so the following discussion is restricted to the mechanical aspects of the machines.

1.2 THE MAIN COMPONENTS OF ROTODYNAMIC PUMPS

1.2.1 A single stage end-suction centrifugal pump

This type of machine is the 'workhorse' design, and is usually directly coupled to an electric motor driven at 1450 and 2900 r/min (50 Hz supply) or 1750 and 3500 r/min (60 Hz supply).

A typical modern design is shown in Fig. 1.1. The pump consists of three components, the liquid end, consisting of the impeller and casing, the seal system, and the bearing housing plus shaft. The wear or neck rings are provided

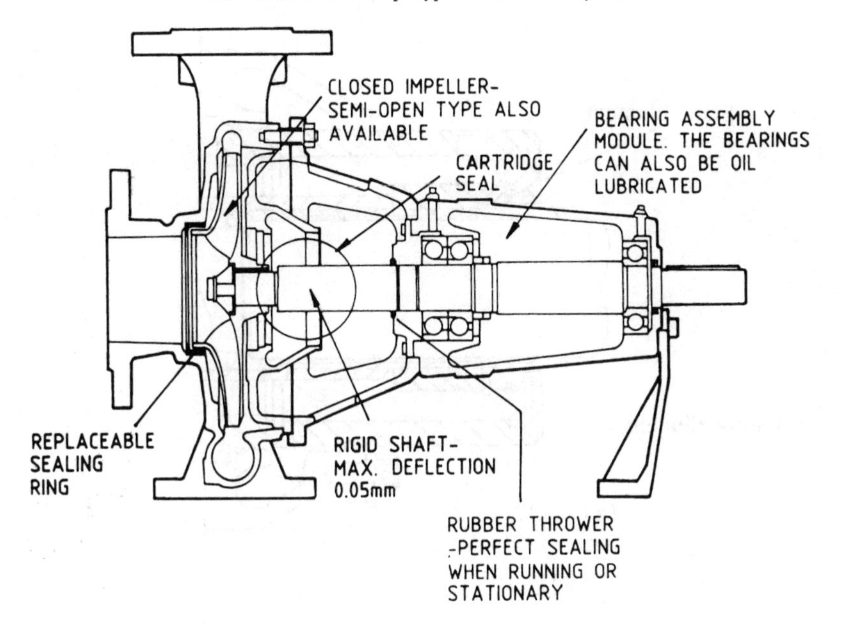

Fig. 1.1 An end-suction, back-pull-out, centrifugal pump

to limit flow back to the suction zone from the discharge zone, and thus maintain the volumetric efficiency. They are of a material which will form a non-galling combination with the impeller, so that if metal to metal contact occurs the impeller will not weld itself to the casing. As will be discussed in Chapter 2, the impeller provides energy input to the fluid. The casing collects fluid from the impeller and in the process of delivering it to the discharge (Fig. 1.3) contrives to reduce fluid velocity and thus provide an increase in static pressure. A typical pump characteristic is shown in Fig. 1.4.

Two types of seal are offered for these machines, a mechanical seal or a packed gland (called a 'stuffing box' by some). A detailed discussion of these systems is beyond the scope of this book, and reference should be made to the IMechE monograph on mechanical seals (**1**) and such texts as the *Seals and Sealing Handbook* (**2**) as well as seal manufacturers' literature. The function of a seal is to ensure liquid is retained in the pump and air is excluded, for centrifugal

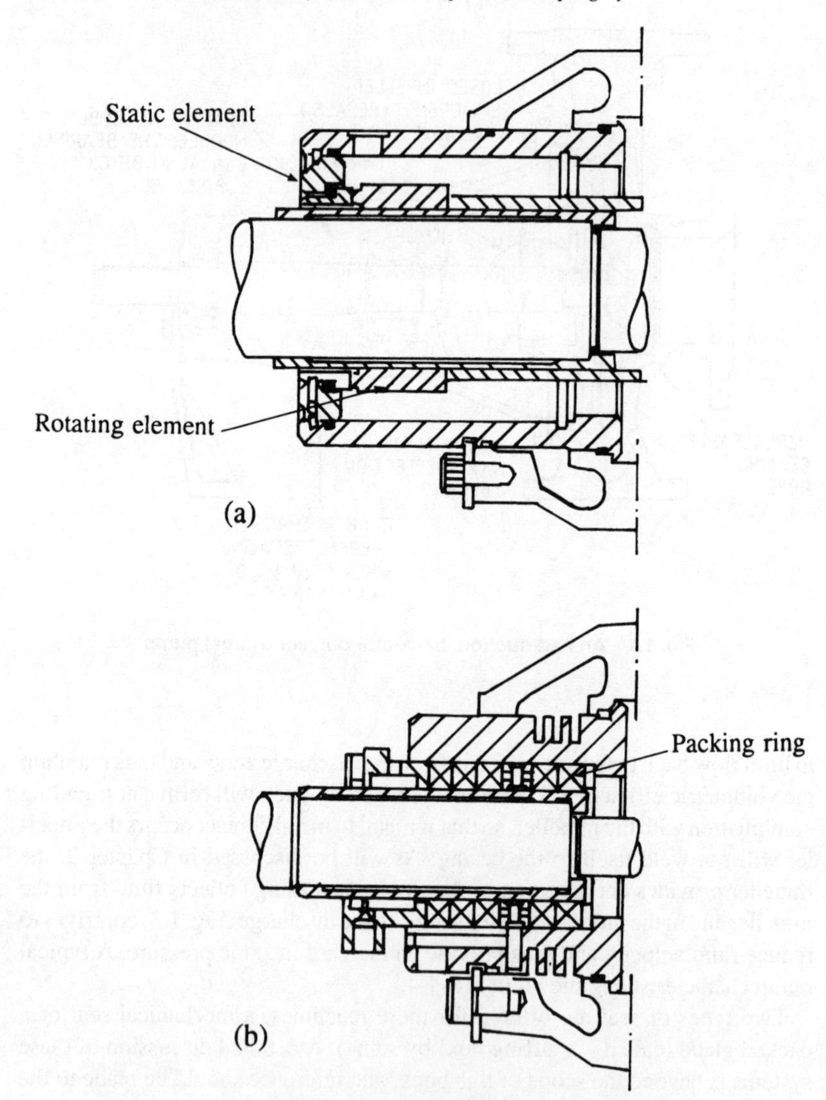

Fig. 1.2 Alternative seal assemblies fitted to the pump in Fig. 1.1
(a) a single mechanical seal
(b) a packed gland

pumps have difficulty in coping with more than *10 per cent* of air in the pumpage. Figure 1.2 shows the modern method of providing cartridge seals of both types.

The bearing housing provides support for the shaft through the bearings. The shaft and bearing system must provide location and withstand the dynamic and hydraulic loads shown in Fig. 1.5.

1.2.2 The double suction pump

This design is provided with a double sided impeller to give twice the flow rate of the end-suction designs. It requires ducting from the suction flange to supply each side of the impeller equally. Figure 1.6 shows a typical design.

1.2.3 The axial flow pump

Figure 1.7 illustrates a simple single-stage axial flow pump, mounted vertically. The main components are the impeller mounted on a shaft system, with the motor mounted above the maximum top liquid level. The casing is approximately cylindrical and is sometimes provided with vanes downstream (called

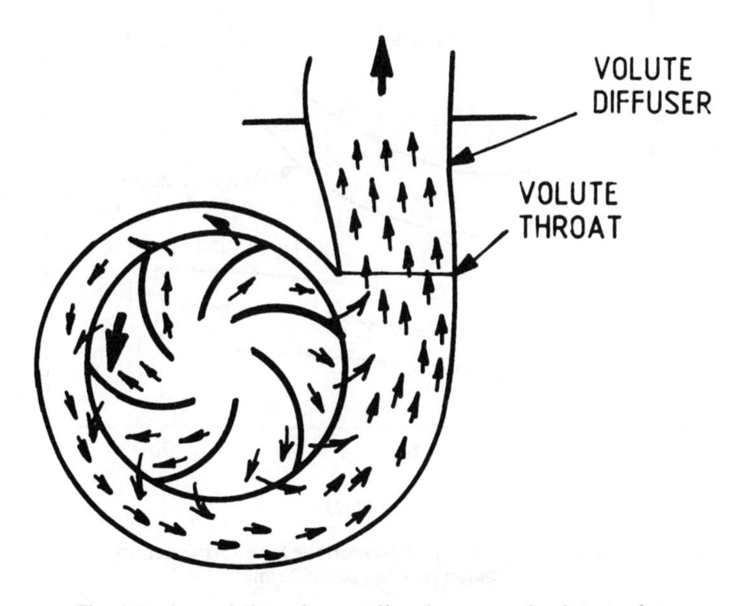

Fig. 1.3 An end view of a centrifugal pump and volute casing

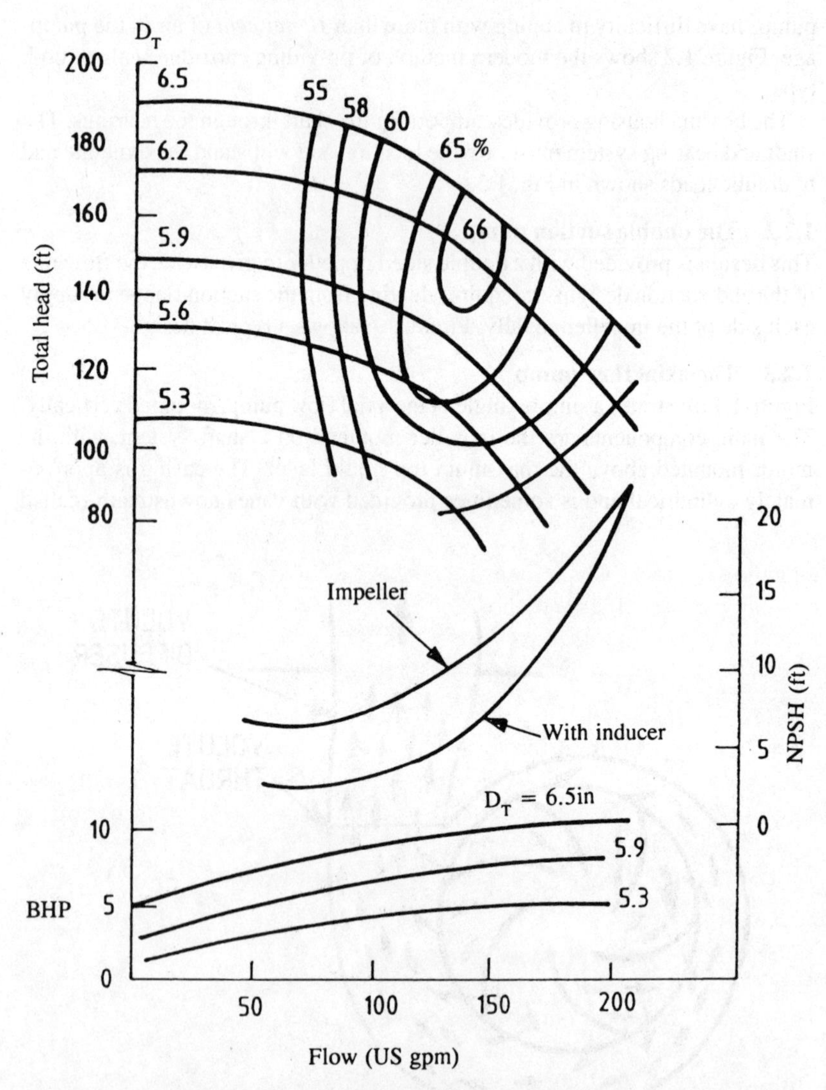

Fig. 1.4 A typical centrifugal pump characteristic for a 2.5 in suction, 1.5 in delivery design running at 3550 r/min

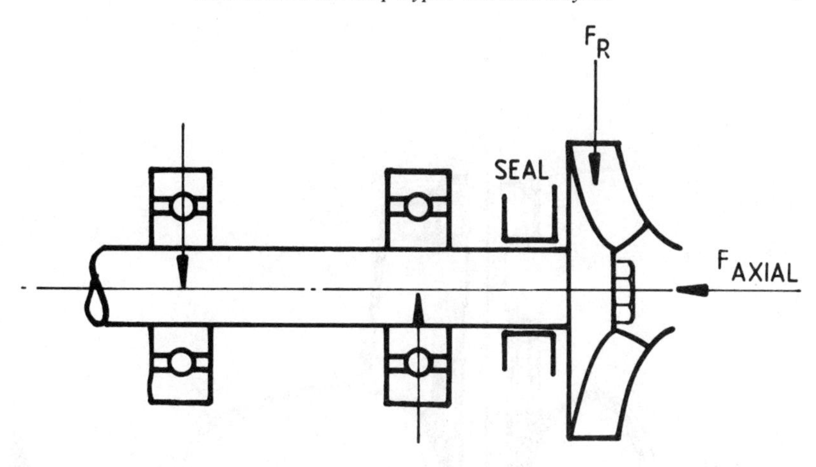

Fig. 1.5 A sketch of a centrifugal pump shaft and bearing system, with hydraulic loads superimposed

outlet guide vanes) which remove impeller-induced swirl and provide some pressure recovery. The same choice of seals as for centrifugal pumps are fitted, but often at the driver end to seal the liquid into the casing, with product-lubricated bearings at the pump end. Such pumps are low pressure-rise and high flow-rate machines.

1.2.4 The mixed flow pump

This design is mixed flow, as sketched in Fig. 1.8, with a volute casing as a centrifugal pump, but many are used as bore-hole pumps (Fig. 1.9), sometimes called bowl pumps. They are mounted vertically, and are either shaft driven as the axial pump in Fig. 1.8, or provided with a special motor which is close connected to the pump. The number of stages depends on the depth of the bore-hole.

1.2.5 Multi-stage centrifugal pumps

These are used for high pressure duties such as boiler feed pump duties. As Fig. 1.10 shows, a number of impellers are mounted in series, each being provided with a diffuser ring and return duct to guide liquid back into the suction zone of the next impeller. For some duties the static components are clamped by straining bolts to retain the internal pressure (as in Fig. 1.10). In modern thermal power station designs the pump stages are located in a pressure casing to provide integrity as a pressure vessel.

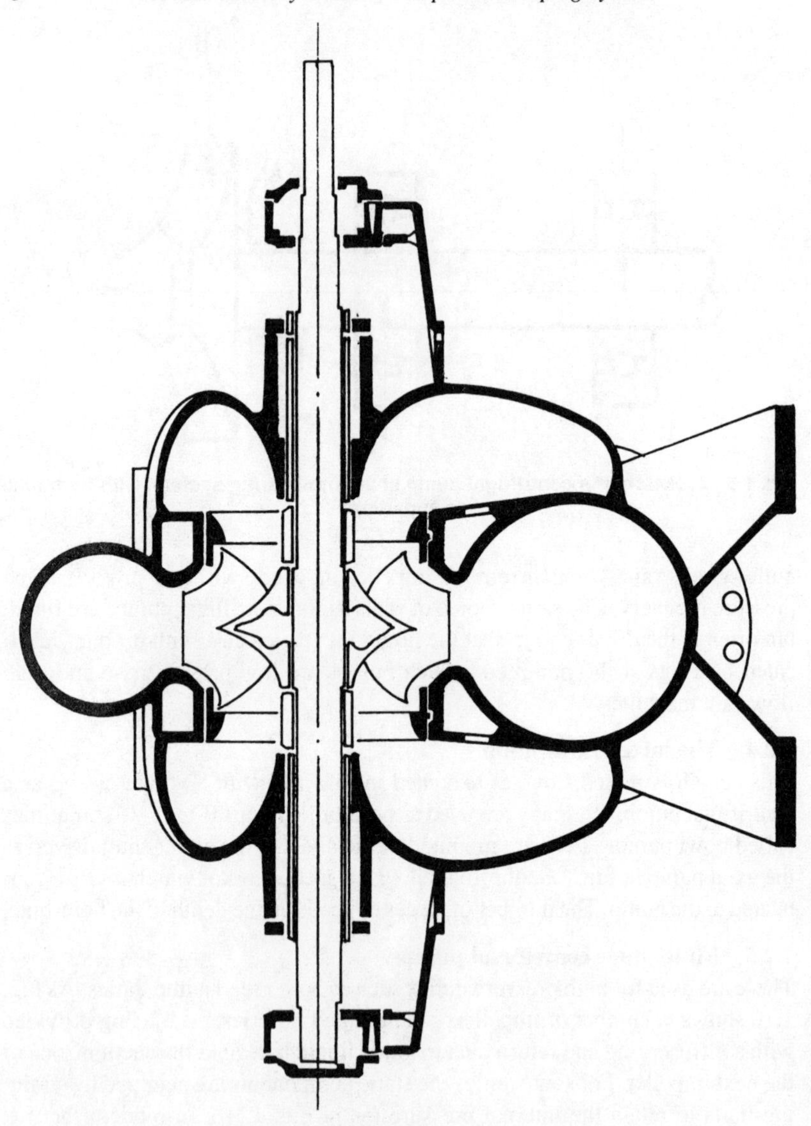

Fig. 1.6 A double suction design

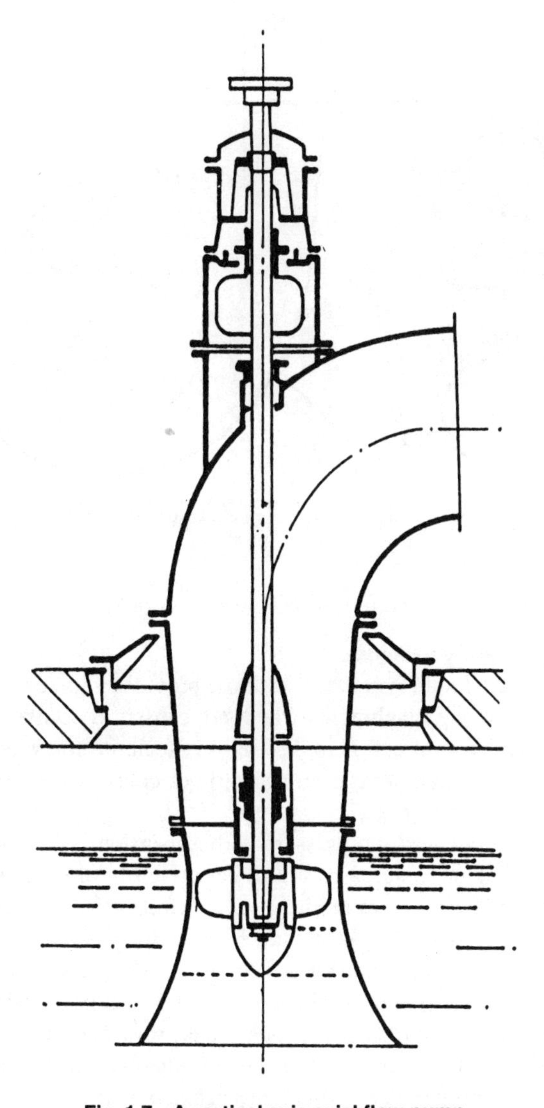

Fig. 1.7 A vertical axis axial flow pump

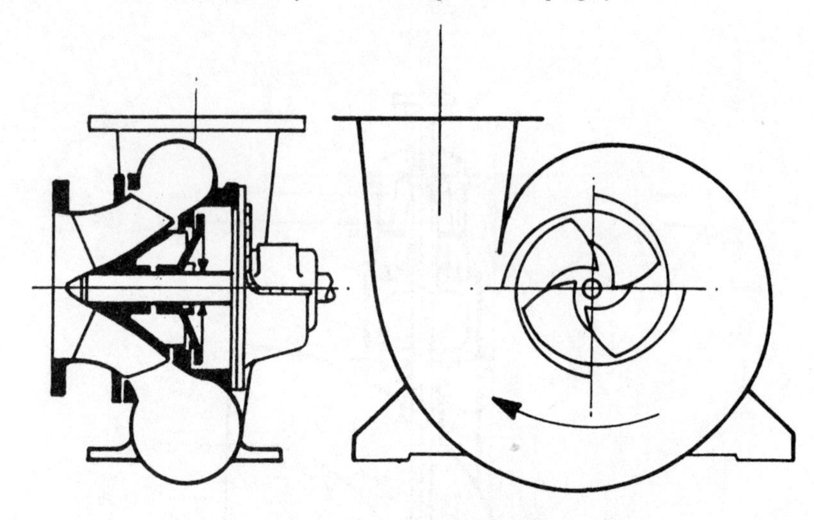

Fig. 1.8 A mixed flow pump

1.3 POSITIVE DISPLACEMENT PUMPS

1.3.1 Reciprocating designs

There are two main families of machine, those provided with reciprocating pistons, and those having diaphragms which are caused to oscillate either by a piston or hydraulic pressure. Figure 1.11 shows a reciprocator, with the components labelled. The usual drive is an eccentric or crank, and flow rate can be altered by changing the stroke.

Pumps used to deliver high pressure and suspensions of solids are called power pumps, and in some cases are direct-driven by double acting steam cylinders with valve designs as in Fig. 1.12 and strokes that vary and thus vary flow rate. Precision in delivery is attained by metering pumps, where dead, unswept, volume is minimized and clearances are carefully controlled. In power hydraulics multi-piston pumps, as shown in Fig. 1.13, are provided with a rotating valve port plate or a rotating cylinder bank; flow control is provided by adjusting the swash plate angle. A diaphragm pump is sketched in Fig. 1.14. Since the delivery pressure of positive displacement machines is a function of system resistance, relief valves must be provided, either in the valve head or the delivery line.

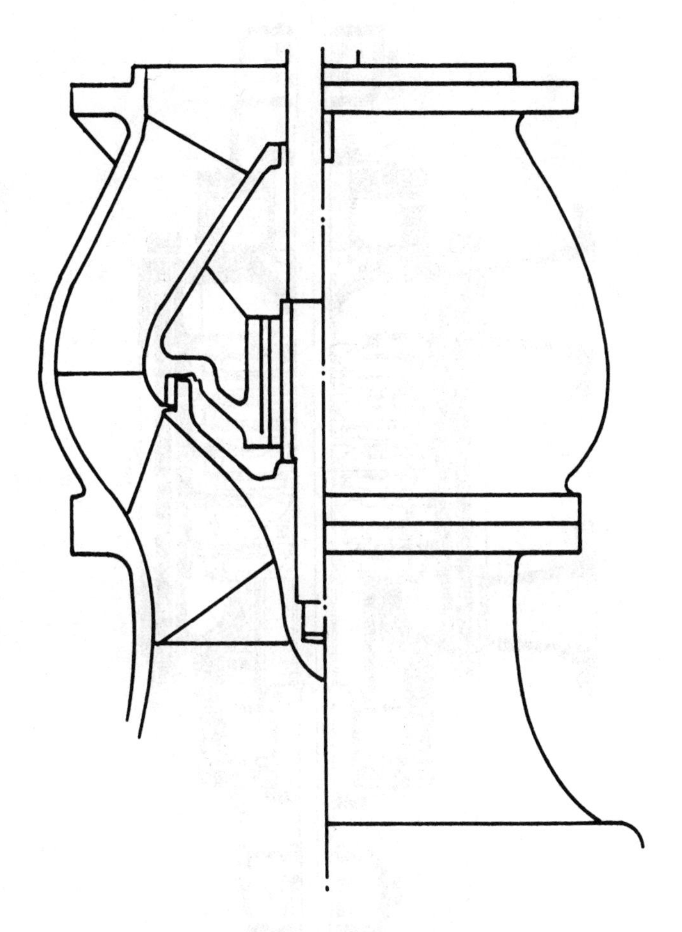

Fig. 1.9 A bowl pump

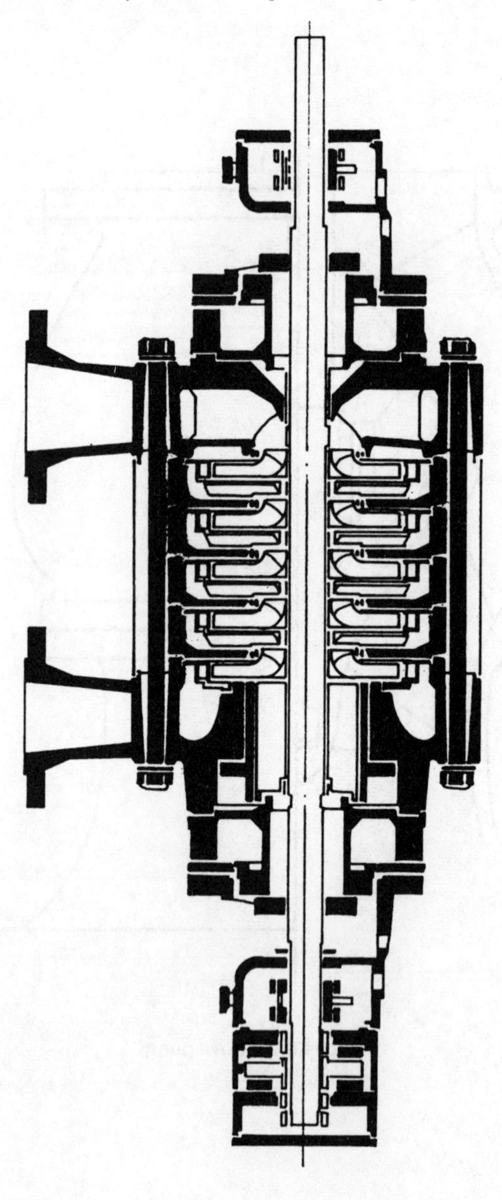

Fig. 1.10 A multi-stage pump

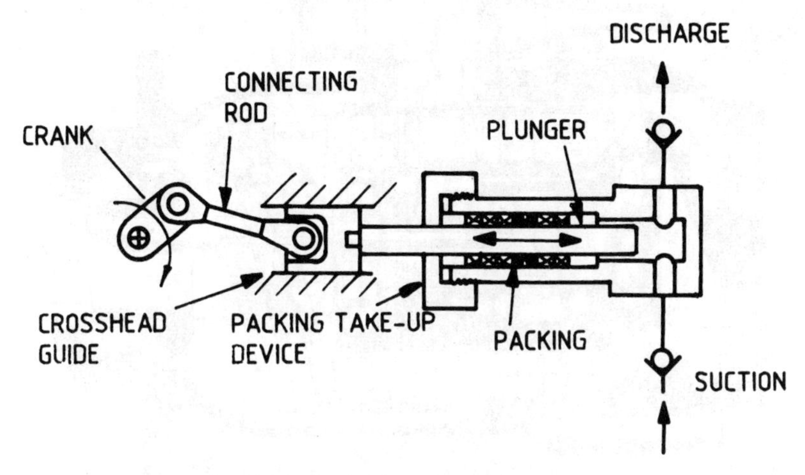

Fig. 1.11 A simple reciprocating pump

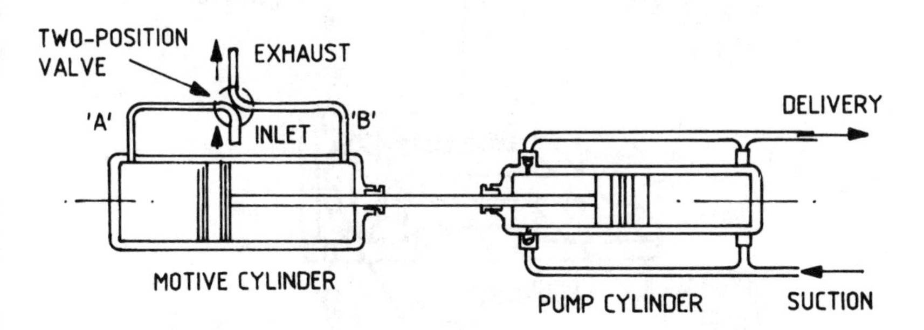

Fig. 1.12 A steam driven double acting reciprocating pump

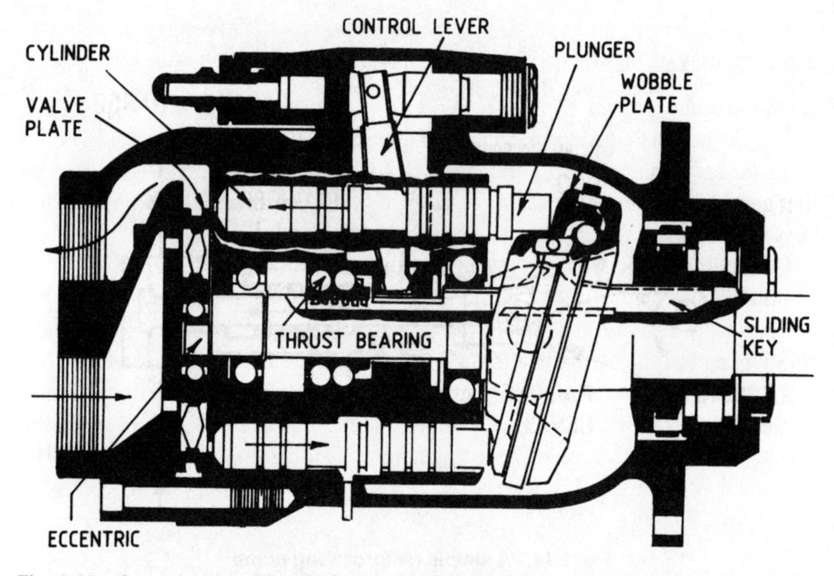

Fig. 1.13 A modern multi-cylinder reciprocating pump using a swash or wobble plate to achieve flow control

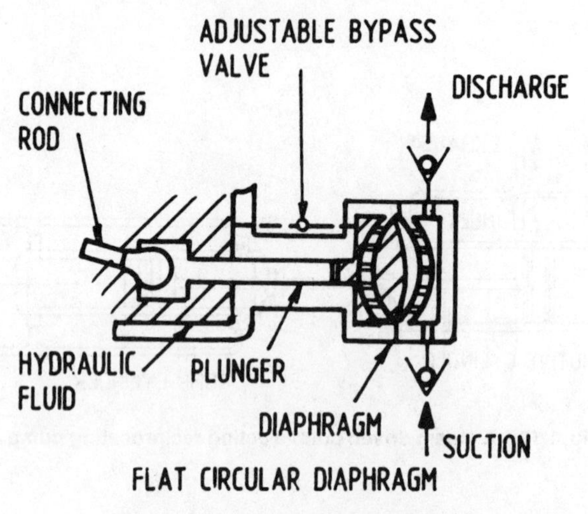

Fig. 1.14 A simple diaphragm pump

In common with all positive displacement pumps, flow control can also be achieved by varying the driving speed, in addition to stroke control.

1.3.2 Rotary pumps

There are four main groups of rotary designs, the Gear and Lobe pumps, Sliding Vane pumps, Screw pumps, and the eccentric rotor Moineau type. A fifth type that can loosely be called rotary is the Peristaltic machine. These will be discussed in order.

Gear pumps consist of two meshing gear wheels of identical size (Fig. 1.15). Liquid is picked up from the suction port by the gear teeth and delivered to the delivery port in fixed 'packets'. Pressure is increased by the dynamic action of rotation, and a typical characteristic is shown in Fig. 1.16. The simplest type has a driver and a follower with contact in the centre which prevents flow back to suction. Precision machines have a set of external gears which provide the mechanical drive and maintain a fixed fine clearance in the meshing zone to provide leakage control. Lobe pumps (Fig. 1.17) are identical in concept with

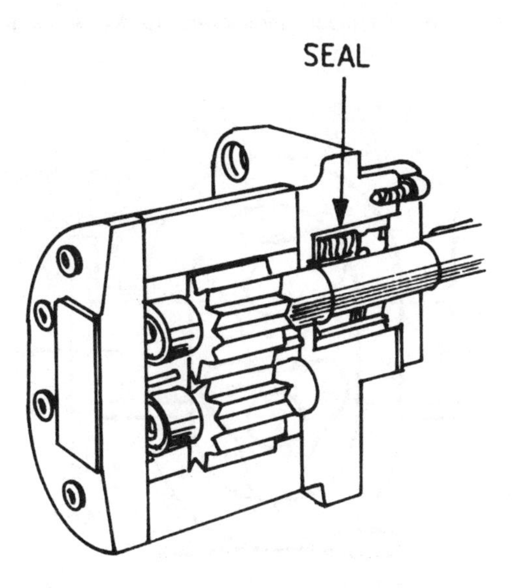

Fig. 1.15 A gear pump

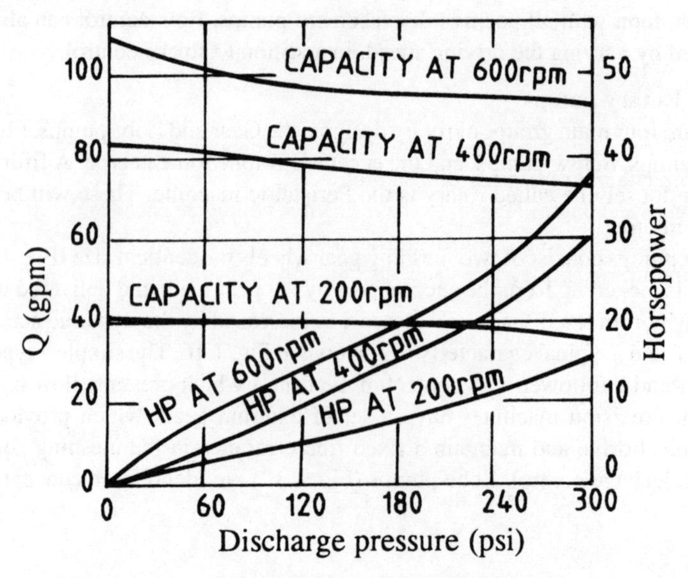

Fig. 1.16 A typical characteristic for a gear pump

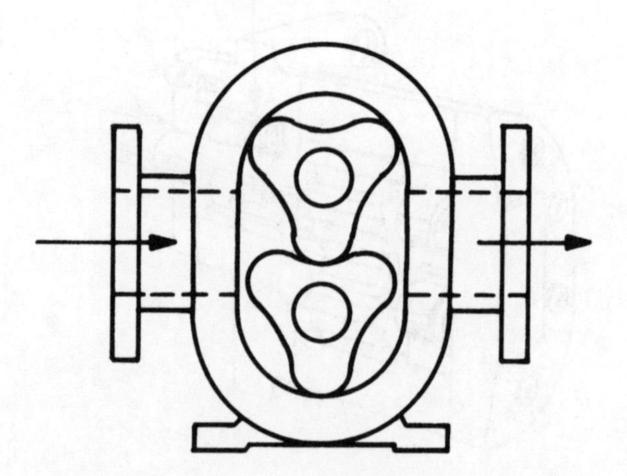

Fig. 1.17 A lobe type rotary pump

the precision gear designs, as the lobes are designed to maintain a constant clearance between the rotor surfaces. These are used extensively for pharmaceutical duties and food processing applications where sterility is important. They are easily steam purged, with no dead spaces, and are designed to be taken apart easily. The 'timing' drive gears are in separate chambers, with lubrication, so the pump chamber needs careful sealing.

Sliding vane pumps (Fig. 1.18) have casings which are eccentric with respect to the rotor, so that vanes slide in and out to form pockets of varying size and thus have a pumping action. The vane, casing, and rotor materials have to be carefully selected to avoid contact welding and to be compatible with the fluid pumped. Vane, gear, and lobe pumps are limited capacity machines.

Fig. 1.18 A simple vane type rotary pump

Screw pumps can deliver pressures and flow rates similar to centrifugal pumps, and are precision machines. Figure 1.19 shows such a machine. The rotors are carefully matched as a set, with very close clearances to give a high volumetric efficiency. Drive is provided by a train of precision timing gears.

The Moineau type of pump has an eccentric rotor driven by a cardan shaft, Figure 1.20. It is provided with a special single helical rotor which runs with an interference fit in a deformable stator which has a double helix form. The rotor axis follows a circular path, hence the need for a cardan shaft which can cope with the eccentricity. The pitch of the stator is twice that of the rotor, and is made from natural rubber, nitrile rubber, or other deformable materials depending on the fluid, and the rotor material can be nitralloy, chrome plated steel, or stainless

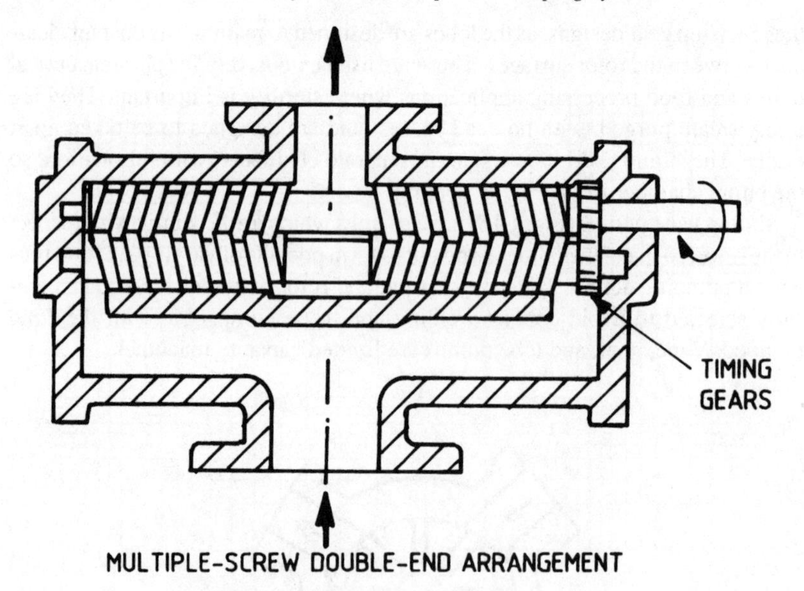

Fig. 1.19 A simple two-rotor screw pump

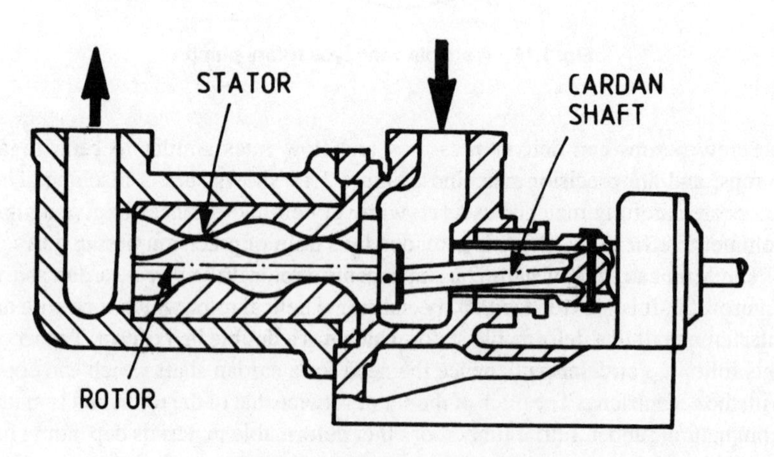

Fig. 1.20 A Moineau type pump

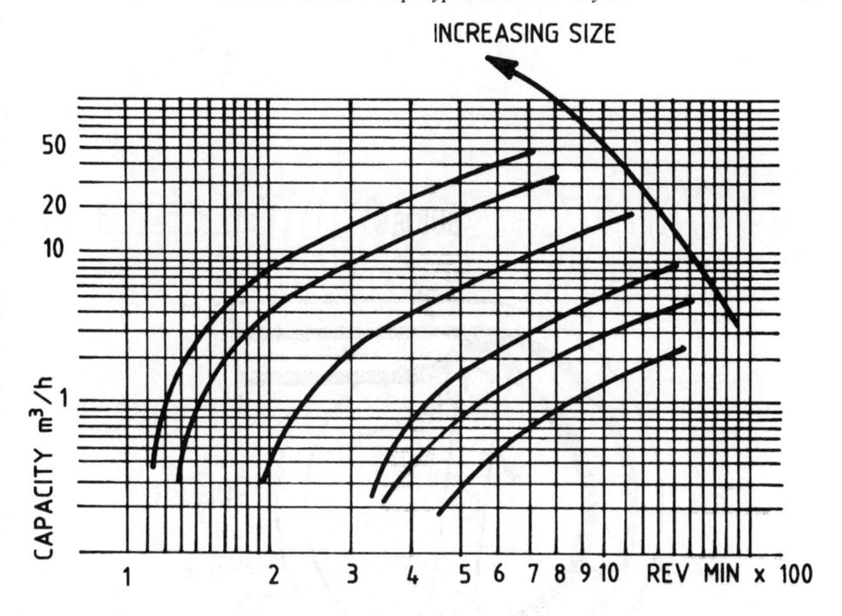

Fig. 1.21 A typical characteristic for a Moineau pump for a 10 bar pressure rise

steel, depending on the duty. This design should never be run dry. The pressure rise will typically be up to 5 bar for a single stage and 10 bar for a two stage design, with flow rates up to 50m^3/hour of clean water, with speed control giving a flow range; Fig. 1.21 is a typical characteristic. This design, being self-sealing needs no inlet or discharge valves.

The Peristaltic machine employs a different principle, as Fig. 1.22 illustrates. A flexible tube is clamped in a semi-circular guide and rollers or shoes press on the tube to close it (or nip it) to seal a section of fixed volume. As the leading roller releases the tube the fixed quantity is discharged. The pressure generated is what is required to overcome the system resistance. This pump is much used in artificial kidney and heart machines, as well as in small dosing systems like drip feeds and water treatment plant. The tubing is replaced when the patient changes in medical use, and as needed by the duty in general use. Recently large designs, to pump gravel and other suspensions, have been produced. This design does not need valves, being self-sealing.

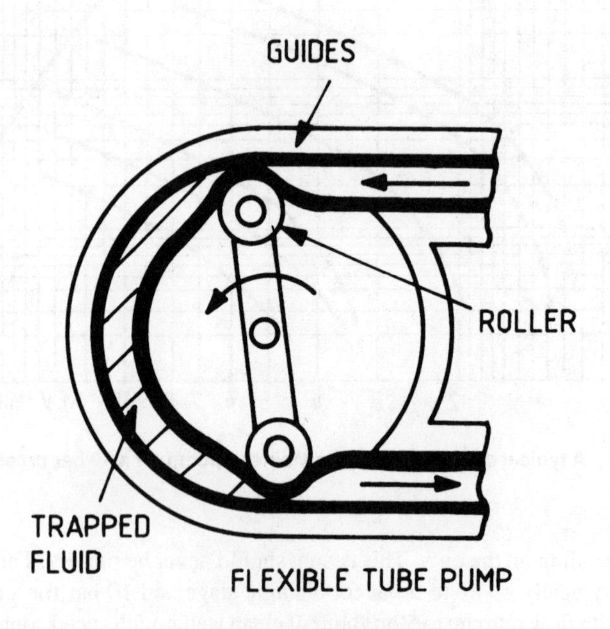

Fig. 1.22 A peristaltic pump

1.4 RANGE CHARTS

To provide a basis for comparison of the main machines described, Figs. 1.23 and 1.24 give the typical range of duties when pumping clean water. As will be discussed in later chapters, fluid properties and cavitation impose limitations, so care is needed if fluids other than clean water are to be pumped.

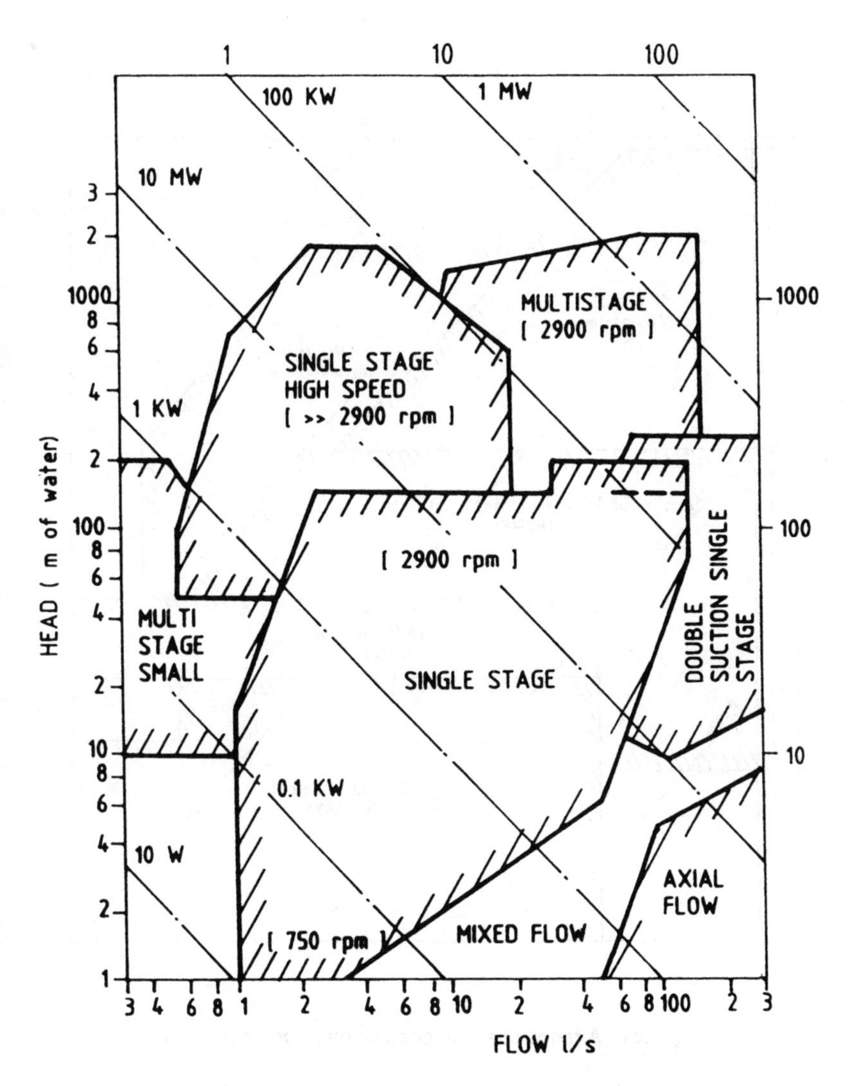

Fig. 1.23 A range chart for rotodynamic pumps

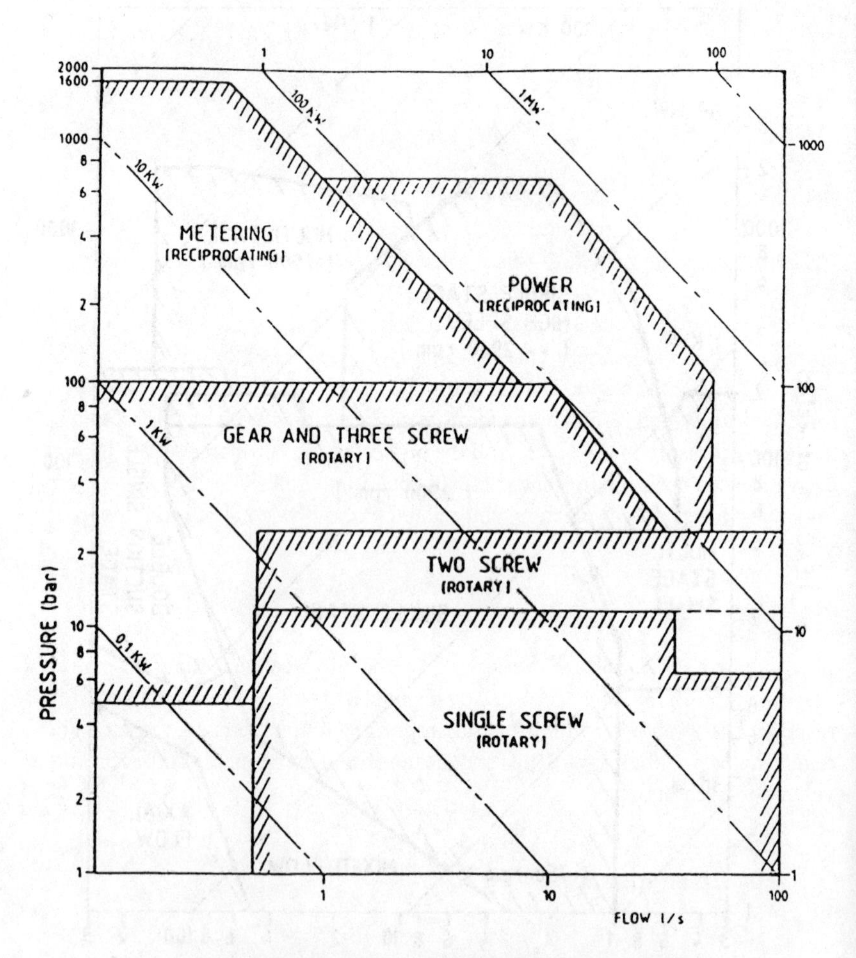

Fig. 1.24 A range chart for positive displacement pumps

CHAPTER 2

Rotodynamic Pump Principles

2.1 INTRODUCTION

Rotodynamic pumps all act continuously on the liquid being pumped, and are usually described in terms of their flow path (Fig. 2.1).

The basic equation used to relate the rotational speed, flow rate, geometry, and energy increase is *the Euler equation*. This was developed to relate the change in angular momentum experienced by the liquid to the energy increase produced by the machine involved. A simple example is the coffee cup and spoon (Fig. 2.2). The spoon, as it is rotated, is a simple impeller, it causes a forced vortex which results in the familiar surface curvature which is related to the lowest pressure in the centre and the highest at the cup surface. The formal derivation can be found in standard texts; the Euler equation will be stated here for a simple centrifugal machine and related to the fluid velocity triangles.

2.2 THE EULER EQUATION

In a simple centrifugal machine (Fig 2.3), the fluid velocity diagram can be drawn for two points, 1 and 2, placed on the inlet and outlet edges of the impeller, respectively, as shown in Fig. 2.4. It can be shown that the energy rise can be related to the peripheral velocities and the tangential components of the absolute velocities at points 1 and 2.

The Euler equation

$$gH_E = U_2 V_{u2} - U_1 V_{u1}$$

or, by using Pythagoras on the triangles

$$gH_E = \tfrac{1}{2}\{(V_2^2 - V_1^2) + \underbrace{(U_2^2 - U_1^2) + (W_1^2 - W_2^2)}_{\text{Static change}}\}$$

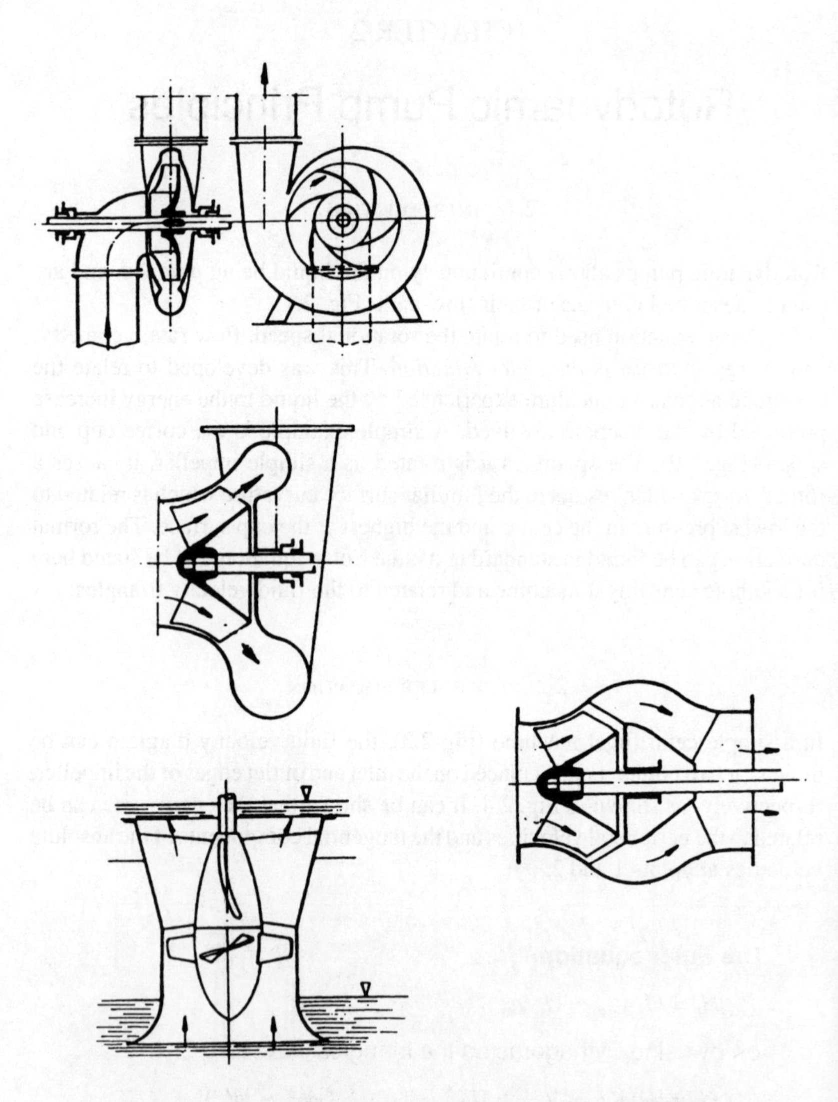

Fig. 2.1 Typical flow paths used in rotodynamic machines

Fig. 2.2 The cup and spoon model

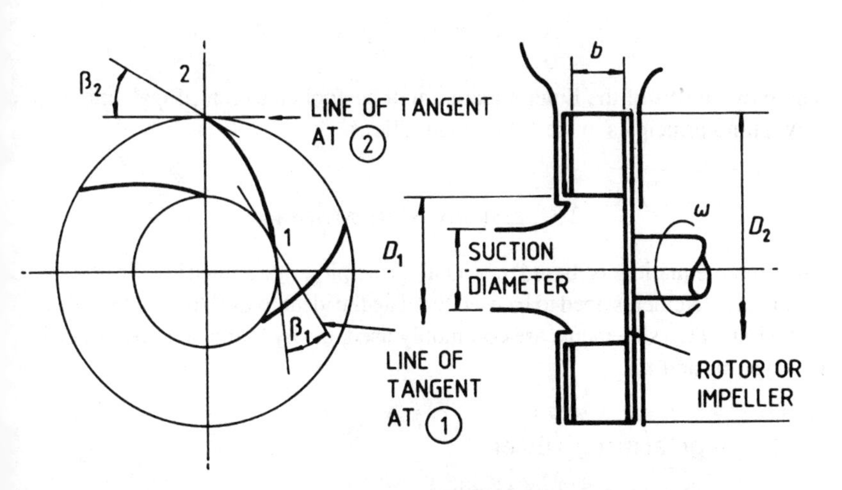

Fig. 2.3 A simple centrifugal pump

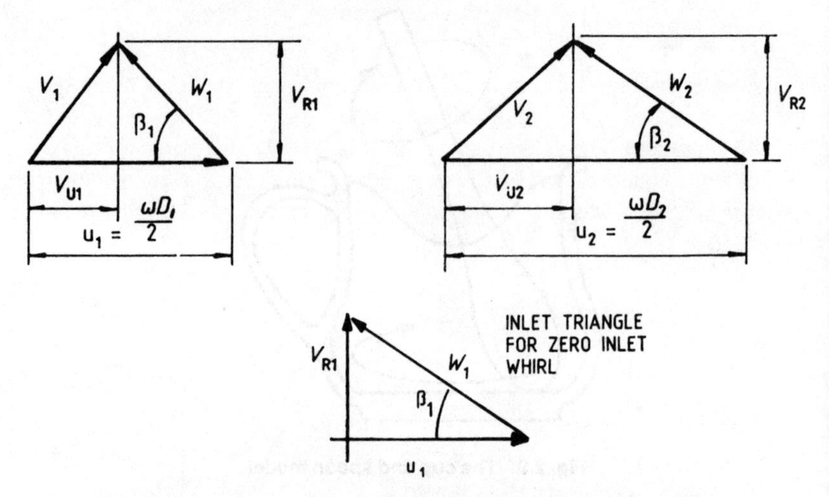

Fig. 2.4 Inlet and outlet velocity diagrams for the pump in fig. 2.3

These two forms of the Euler equation will be applied to centrifugal and axial flow pump principles in the following sections.

2.3 EFFICIENCY DEFINITIONS

The Euler equation is based on ideal principles, ignoring flow losses, so efficiency statements need to be used to relate the ideal 'work' to the actual fluid behaviour. Two statements are commonly used, the hydraulic and overall efficiencies, η_H and η_O

$$\eta_H = gH \text{ actual}/gH \text{ Euler}$$

$$\eta_O = \dot{m}\, gH \text{ actual/Input power}$$

2.4 CENTRIFUGAL PUMP PRINCIPLES

If the pump discussed in section 2.2 is considered again, the Euler equation can be further refined using the assumption of zero inlet whirl (often called the design condition). The velocity diagrams (Fig. 2.4) can be drawn for this case.

The Euler equation can be written without the inlet component, and then re-written to give a working ideal equation for the simple centrifugal pump considered.

$$gH_E = U_2 V_{u2}$$

since in Fig. 2.4

$$V_{U2} = U_2 - V_{R2} \cot \beta_2$$
$$gH_E = U_2^2 - (QU_2 \cot \beta_2)/A_2$$

since

$$V_{R2} = Q/A_2$$

where

$$A_2 = \pi D_2 b$$

The ideal specific energy rise is, therefore, a function of U_2 and flow rate for a given size of machine, as is illustrated in Fig. 2.5.

Also shown in the figure is an actual curve which shows how pump flow losses reduce the energy rise below that for the ideal flow predicted by the Euler equation.

2.5 AXIAL FLOW PUMP PRINCIPLES

For a simple axial flow pump consisting of a single rotor sketched in Figure 2.6 if the ideal zero inlet whirl condition is assumed.

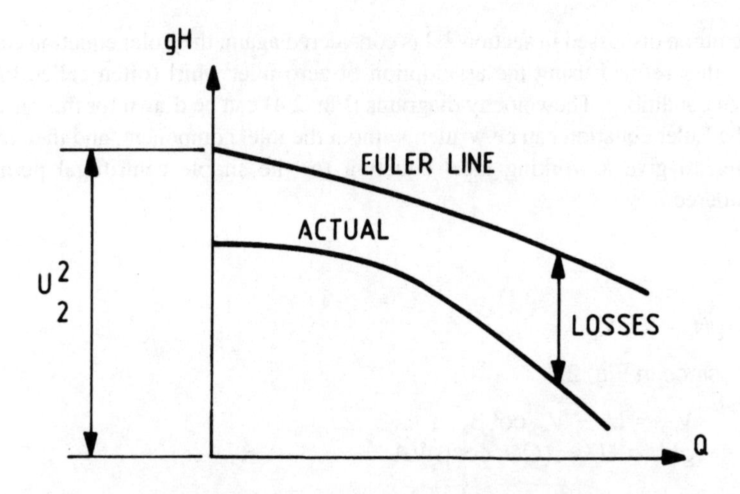

Fig. 2.5 The ideal 'Euler' and actual pump characteristics

$$gH_E = U\,V_{u2}$$

or

$$gH_E = \tfrac{1}{2}\{(V_2^2 - V_1^2) + (W_1^2 - W_2^2)\}$$ — **The axial form of the Euler equation**

This is the simplest form of equation for the Ideal condition. If a similar exercise is followed as for the centrifugal pump

$$gH_E = U^2 - \frac{4Q\,U\cot\beta_2}{(D_T^2 - D_H^2)}$$

Which is a simple relation of Euler energy change to geometry.

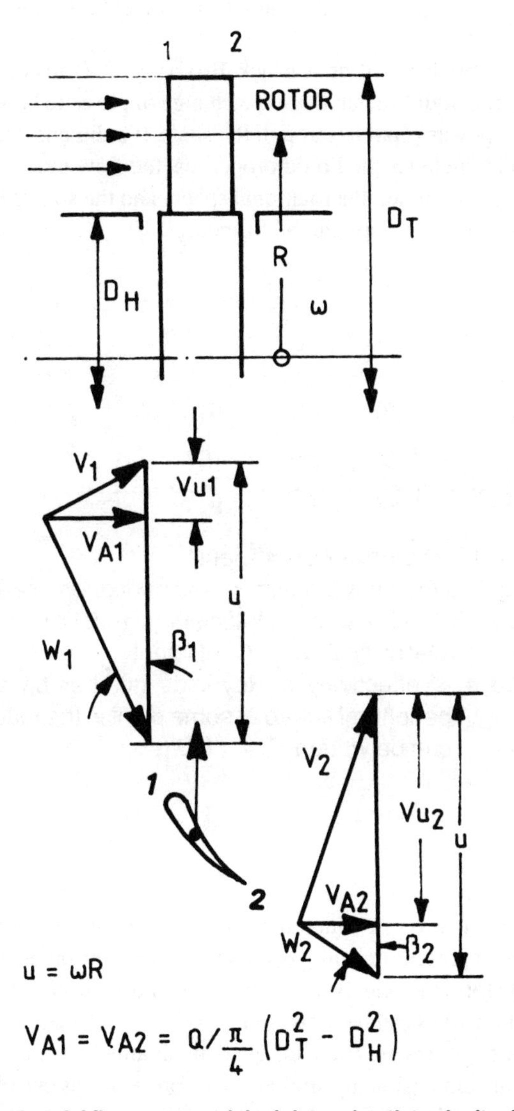

Fig. 2.6 A simple axial flow pump and the inlet and outlet velocity diagrams

2.6 SIMILARITY, MODEL LAWS, SPECIFIC SPEED

A pump can be depicted as a Black Box (Fig. 2.7) inside which energy is imparted to the liquid so that there is a change of pressure between the inlet and outlet ports; power is put in through the shaft. If a dimensional analysis is performed which includes the liquid properties, the flow rate, the specific energy change, the power input, the rotational speed, and the size of the machine, a set of non-dimensional groups can be formed.

(1) (2) (3) (4)

$$\frac{P}{\rho\omega^3 D^5} = f\left(\frac{Q}{\omega D^3}, \frac{gH}{\omega^2 D^2}, \frac{\rho\omega D^2}{\mu}, \ldots\right)$$

Group 1 is the Power Coefficient
Group 2 is the Flow Coefficient (symbol often used: ϕ)
Group 3 is the Pressure Coefficient (symbol ψ)
or Specific Energy Coefficient
Group 4 is effectively a Reynolds number based on the peripheral speed at some point in the machine, and can be written $R_e = \rho UD/\mu$

There are two ways in which these groups can be used: one is to provide a rational way of presenting pump performance data; the other is to give a method by which performance can be predicted. If a pump is tested at a given speed, its characteristic curves can be presented in the usual manner, as in Fig. 2.8. If the dimensionless groups are used, all the data obtained for machines in the same family that are dimensionally similar can be presented as one plot, as in Fig. 2.9, rather than a number of figures related to each speed and machine size in the family.

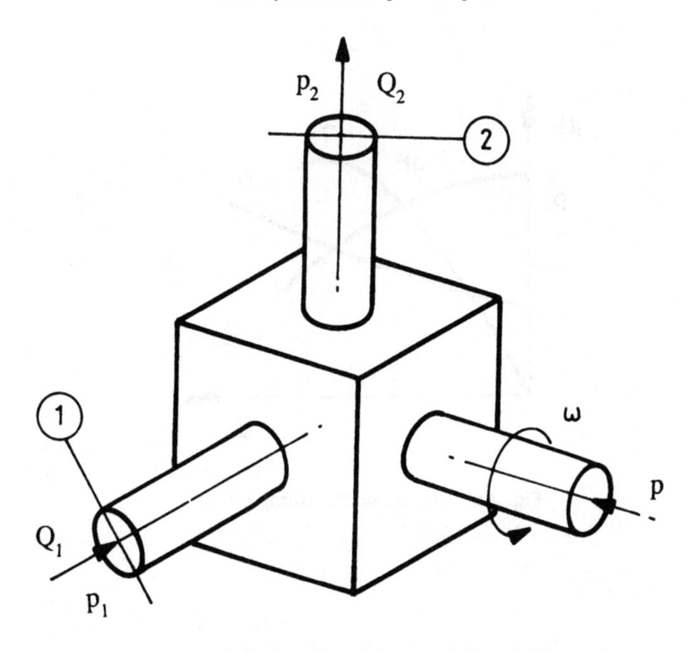

Fig. 2.7 A 'black box' representing a pump

If the rotational speed of a pump is reduced from ω_1 to ω_2, the dimensionless groups may be used to predict the dynamically similar flow conditions that will follow, as illustrated in Fig. 2.10, where

$$\left.\begin{aligned} Q_2 &= Q_1\left(\frac{\omega_2}{\omega_1}\right) \\ gH_2 &= gH_1\left(\frac{\omega_2}{\omega_1}\right)^2 \\ P_2 &= P_1\left(\frac{\omega_2}{\omega_1}\right)^3 \end{aligned}\right\} \text{'Scaling Rules'}$$

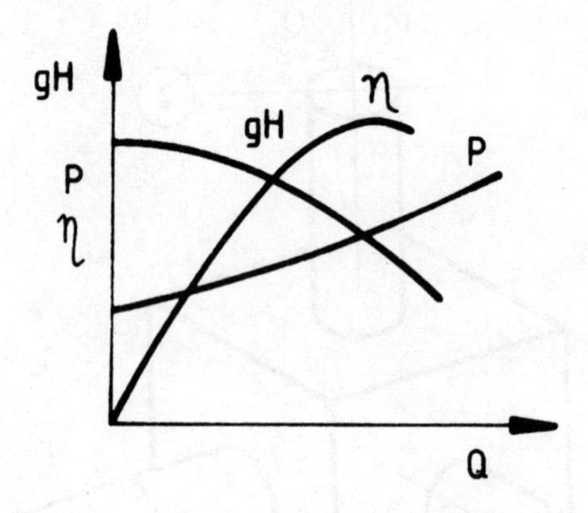

Fig. 2.8 Characteristic pump curves

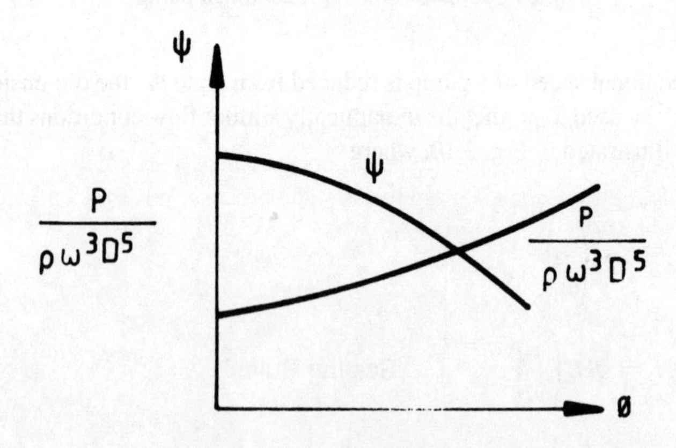

Fig. 2.9 Dimensionless pump curves

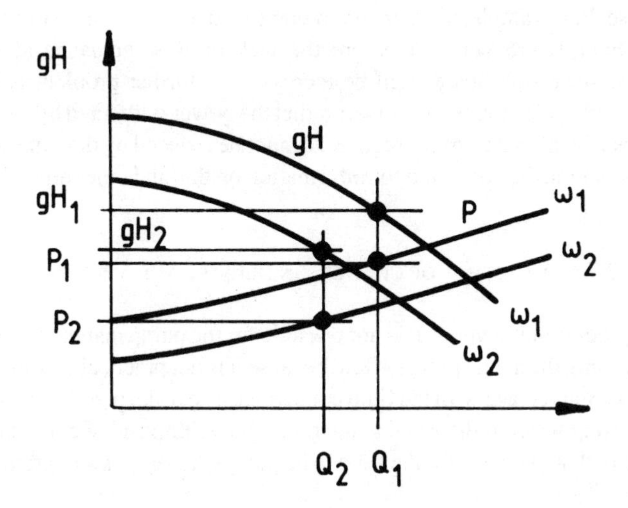

Fig. 2.10 Prediction of the change in performance with speed change using the scaling laws

If the converse situation applies, as often occurs in practice, and the diameter reduction is needed to reduce the pump performance by a small amount, the dimensionless groups may be used to assist the process. What are known as the Scaling Laws are derived from them.

$$D_1 = D_2\sqrt{\left(\frac{H_1}{H_2}\right)};\ Q_1 = Q_2\sqrt{\left(\frac{H_1}{H_2}\right)}$$

$$D_1 = D_2(Q_1/Q_2);\ gH_1 = gH_2\left(\frac{Q_1}{Q_2}\right)^2$$

$$D_1 = D_2\,(gH_1/gH_2)^{3/2}$$

The pump handbooks may be consulted for examples of the way that this set of scaling formulae may be used, and also the reserve that needs to be exercised in their use. For example, the diameter reduction that is considered to be acceptable is about 10–15 per cent before the lack of flow guidance affects the efficiency, since flow losses will be increased. A further problem is that the accuracy with which the scaling laws predict the power is affected by the surface roughness which, as pump size reduces, cannot be reduced by the same amount. Textbooks should be consulted for information on this and other related effects.

2.7 THE EFFECT OF CHANGES IN DENSITY AND VISCOSITY

In many process applications it is not possible for the pump manufacturer to test the pump using the actual process fluid because it is not practicable to drain their water tanks and replace with the fluid concerned. The usual practice is to test the machine using water at the usual temperature prevailing and to correct the performance to that expected for the fluid to be pumped, using a nomogram such as Fig. 2.11.

If the viscosity of the liquid is near to that of water, but the density is different, the dimensionless groups discussed in section 2.6 may be used with some accuracy.

If the viscosity is different to that of water, experience suggests that unless the kinematic viscosity exceeds 20–30cSt correction is not needed. If the kinematic viscosity does exceed that value correction is needed, and a nomogram like the one illustrated in Fig. 2.11 may be used.

2.8 SPECIFIC SPEED OR CHARACTERISTIC NUMBER

The dimensionless groups and the Euler equations that have been discussed all include the size of the machine. One set or group of quantities that have been used for some time are composed of the rotational speed, the flow rate, and the specific energy rise. These groups are called Specific Speeds, and use the design or best efficiency flow rate and specific energy rise.

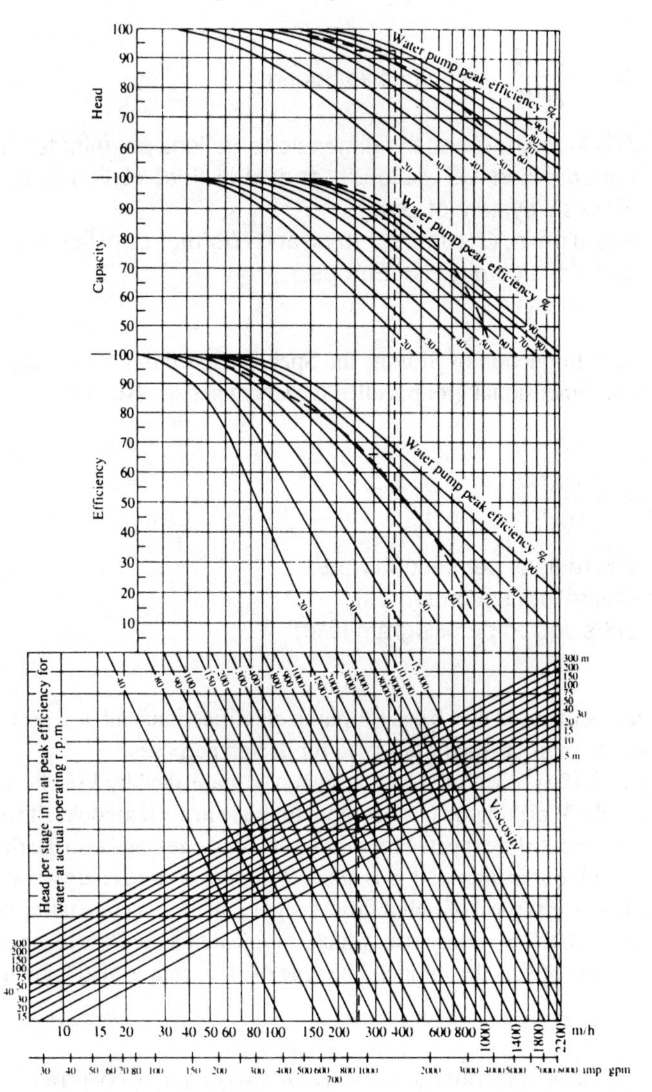

Fig. 2.11 A viscosity correction nomogram based on that quoted by (from Davidson, (3), 1993, *Process Pump Selection – A Systems Approach*, Second Edition, IMechE, London.)

$$N_s = \frac{N\sqrt{Q}}{H^{3/4}} \quad \text{— Specific Speed}$$

N is in r/min, Q has the dimensions 'gallons per minute' (both Imperial and US gallons), and H is 'feet of liquid' in the 'English' system of units.
N is in r/min, Q is l/min, or m^3/s, and H is 'metres of liquid' in the variations common in Europe.

With the introduction of SI units the Specific Speed has been replaced with a truly non-dimensional group, called the Characteristic Number, k_s

$$k_s = \frac{\omega\sqrt{Q}}{(gH)^{3/4}}$$

ω is radians per second (rad/s)
Q is m^3 per second (m^3/s)
gH is Joules per kilogram (J/kg)

This number is used as a basis for pump classification, and as a base line for a number of empirical quantities used for design purposes.

Figure 2.12 is based on a well known plot, published by Wisclicenus **(4)** on behalf of the Worthington Corporation, found in many textbooks. It shows typical efficiencies to be expected for a range of pump sizes, and also the flow path to be expected. For example, if k_s is small it implies a large energy rise for small flow rates, as is typical of radial flow designs, and if k_s is large it implies a large flow rate with a low energy rise, which is typical of axial flow machines. Thus, when a value of k_s is determined, the type of flow path is also known.

2.9 CENTRIFUGAL IMPELLER AND CASING GEOMETRY

The Euler equation indicates that the performance is related to the outer diameter of the impeller, the outlet angle, the rotational speed, and the width of the

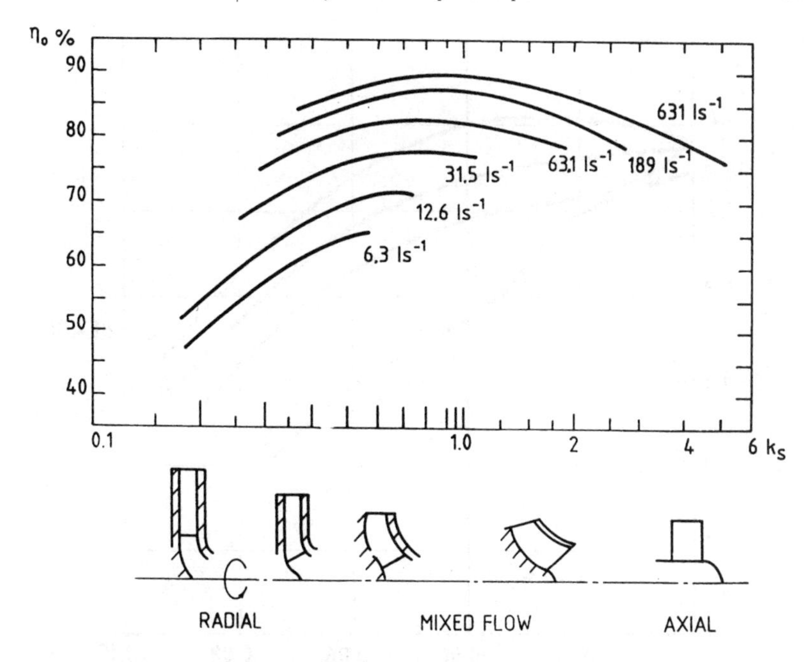

Fig. 2.12 A plot of overall efficiency against varying flow and characteristic number, k_s (based on the well known Worthington Plot given by Wisclicenus (4))

impeller passages. Further factors are the shape of the blade passages and the number of blades. Experimental studies have provided the understanding of the relation between guidance and the need to keep the wetted area to a minimum, established many years ago by Professor Thoma. A large number of blades provides good guidance, but at the expense of losses. Figure 2.13 indicates that for a conventional centrifugal pump there is an optimum number of blades which gives a good head-to-flow curve which rises steadily as flow rate reduces, with the maximum head at zero flow rate, and also gives the best efficiency. In modern designs blade numbers between five and seven are often provided. For mixed flow machines the need is to provide guidance of the strongly three-dimensional flow, so in some cases more blades are used, but in other cases there is too much 'metal' in the suction zone and fewer blades are used. Reference may be made to texts such as Stepannof (6) for more details.

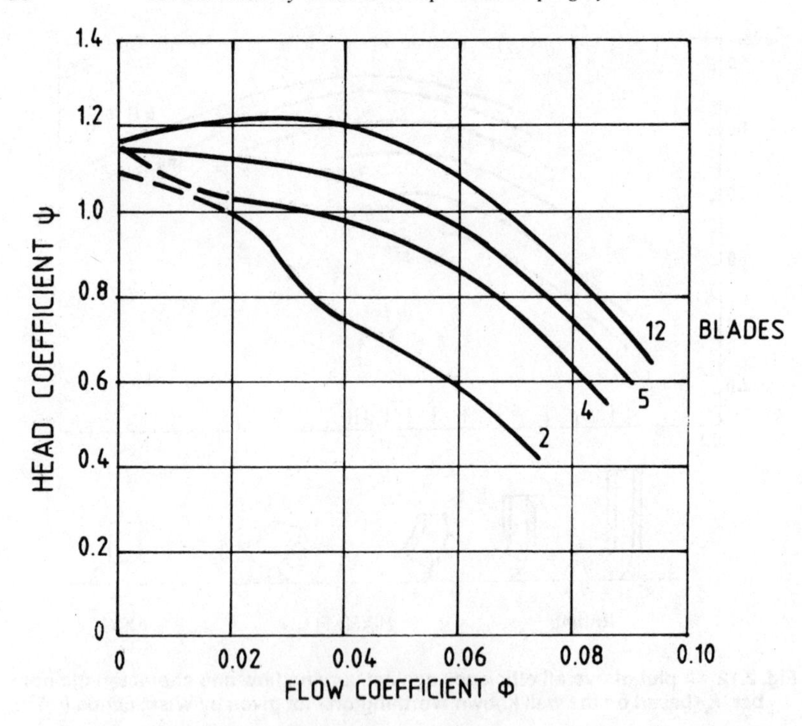

Fig. 2.13 The effect of blade number on centrifugal pump performance (based on the paper by Varley (5))

Having established blade numbers the designer has to produce the blade angles needed at inlet and outlet. It is usual to provide an incidence of about 3 degrees at inlet referred to the ideal fluid angle, as shown in Fig. 2.14, and at outlet to allow for what is called 'slip' by many engineers. This is a correction for lack of guidance at outlet that results in a considerable deviation from the Euler triangle, as shown in Fig. 2.15. A number of formulae are available based on the outlet angle of flow and the machine geometry, and these will be found in the textbooks quoted in the bibliography.

Having decided the blade angles and diameters of the impeller, the blade shapes must be determined. The ideal shape from inlet to outlet is the Archimedean spiral, which is usually approximated on the drawing board by using mul-

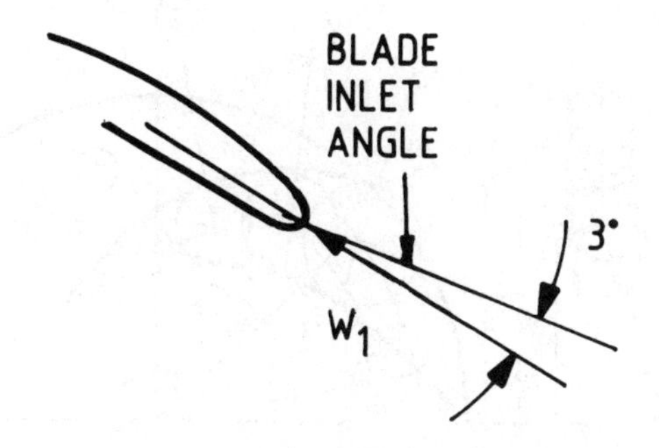

Fig. 2.14 The concept of incidence

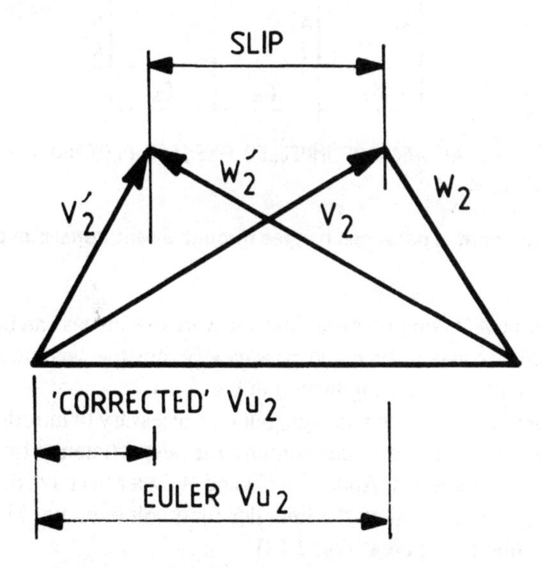

Fig. 2.15 The definition of the slip velocity

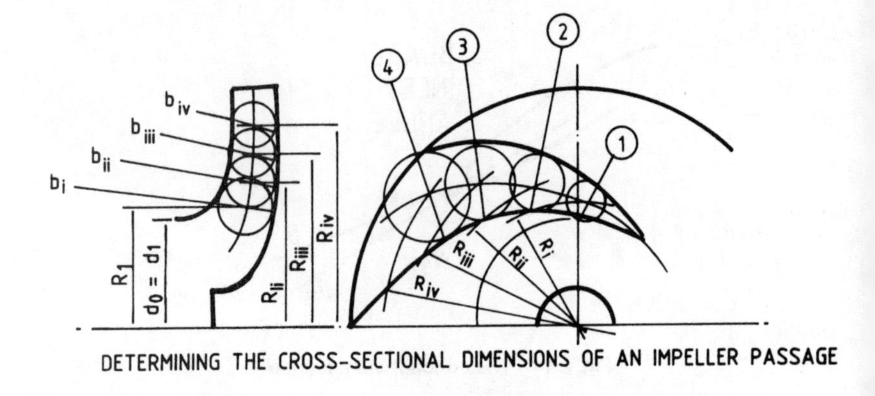

DETERMINING THE CROSS-SECTIONAL DIMENSIONS OF AN IMPELLER PASSAGE

STATIONS ALONG PASSAGE

A1 1 | A2 2 | A3 3 | A4 4

δx δx δx

CROSS-SECTIONAL AREA OF IMPELLER PASSAGE PLOTTED AGAINST PASSAGE LENGTH

Fig. 2.16 The normal passage changes through a centrifugal pump impeller

tiple radii. With CAD computer drawing software the shapes can be accurately plotted. The ideal shapes for blade passages permit the velocities to reduce slowly from inlet to outlet, as indicated in Fig. 2.16.

This is not a treatise on pump design, but it is necessary to introduce a further concept of great importance in determining the best efficiency flow rate for a pump; this is the Area Ratio. Anderson (7) and Worster (8) have discussed this way of relating the casing with the impeller for a volute pump. The important factor is the volute throat area (Fig. 2.17).

The argument is that the volute is effectively a passage which has a rising resistance which, as a first approximation, can be a straight line. Where this meets the Euler impeller line should approximate to the machine match point, as

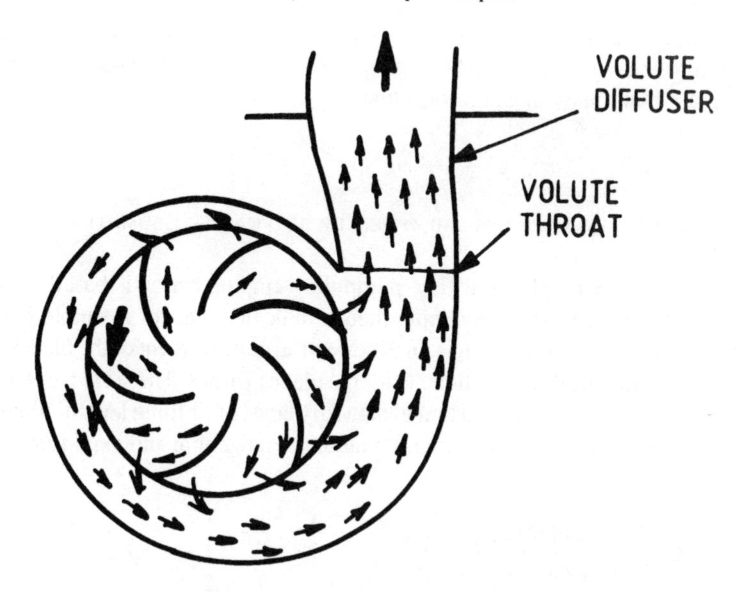

Fig. 2.17 The volute collector casing

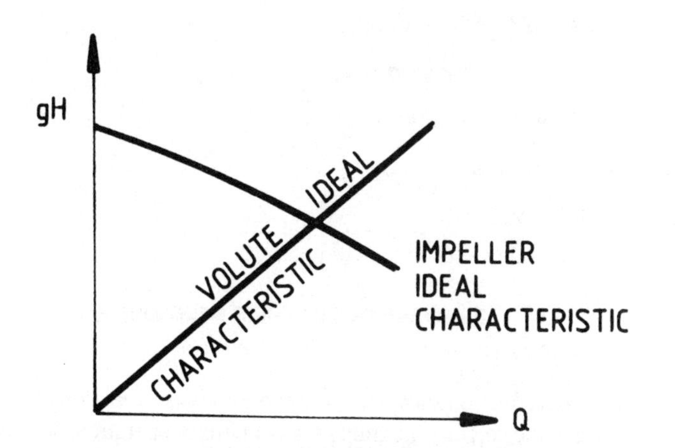

Fig. 2.18 The ideal impeller and volute design diagram proposed by Worster based on an idea by Anderson

shown in Fig. 2.18, which should give the flow rate that will result in the best efficiency behaviour. For details of pump design the books and articles referred to in the bibliography must be consulted.

2.10 AXIAL FLOW PUMP IMPELLER AND CASING GEOMETRY

In the case of the simple axial flow pump the pumping action is based on the action of the blades, and the simple theory indicates that for a simple blade profile the tangential force (that is, in the direction of movement of the blades) is related to the lift and drag coefficients for the blade profiles. These are based, for isolated blade profiles, on wind tunnel data for blades of 'infinite length'. Figure 2.19 illustrates the relation of lift and drag to the tangential and axial forces.

$$C_L = \frac{\text{Lift force } (L)}{\frac{1}{2}\rho V^2 \times (\text{proj. area})}$$

$$C_D = \frac{\text{Drag force } (D)}{\frac{1}{2}\rho V^2 \times (\text{proj. area})}$$

$$F_T = \rho V_A^2 S (\cot \beta_1 - \cot \beta_2)$$

since the blade is moving

work done per second = $F_T \times u$

Blade loading ψ is given by

$$\psi = \frac{gH}{U^2} = \frac{V_A C_L}{2U} \cdot \frac{C}{S} \operatorname{cosec} \beta_m \left[1 + \frac{C_D}{C_L} \cot \beta_m\right]$$

($\psi = gH/U^2$; this is the basic design equation)

The basic design equation shows that the energy rise is a function of the blade lift coefficient and the velocity diagrams at any radius of operation, defined as in Fig. 2.6. The typical blade profile data shown in Fig. 2.20 is presented against angle of attack. The lowest loss case corresponds to the highest value of the lift

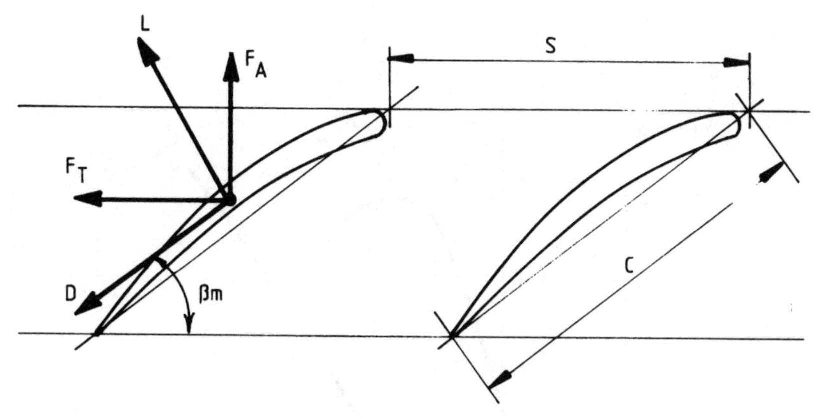

Fig. 2.19 Lift and drag forces on an aerofoil

to drag ratio C_L/D_D and the maximum angle of attack that can be tolerated is that relating to the stall point, where C_L reaches its peak value. The angle of attack is defined by referring flow direction to the line of zero lift which makes an angle, x, with the chord line of the profile, as sketched in Fig. 2.21(a). The line of zero lift is found from test data, but for a cambered blade can simply be constructed for a normal blade profile, as shown in the diagram in Fig. 2.21(b)

Profiles used can be either flat bottomed wing sections or profiles developed for compressor use, such as around a camber line.

The design process starting with the duty and working through the Euler equation to give the blade loading continues with assigning a value of C_L for a chosen profile for a given angle of attack. Mechanical design then follows the determination of blade number, and the chords of the blades, and for this process the textbooks cited may be consulted.

2.11 HYDRAULIC LOADS, RADIAL AND AXIAL THRUST

A centrifugal pump assembly consists of an impeller and casing (the liquid end), a shaft system, a bearing assembly, and a sealing system. The whole shaft and bearing system must be designed to withstand all loads applied to it. Two load

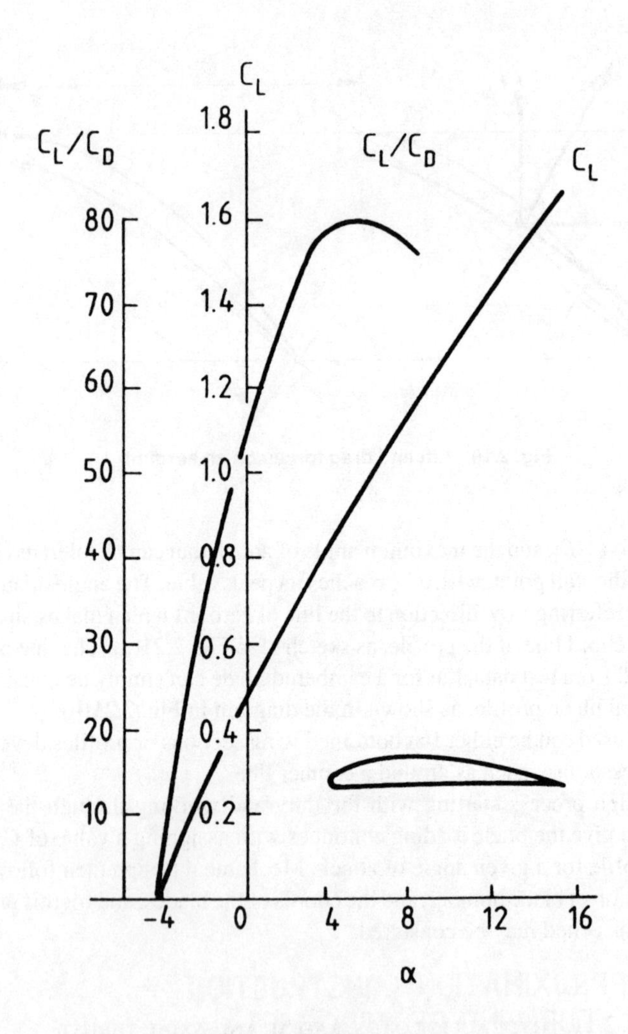

Fig. 2.20 A typical low speed aerofoil characteristic

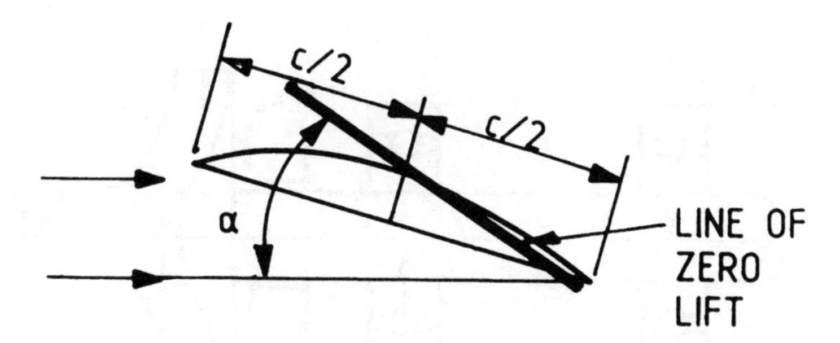

Fig. 2.21(a) Definition of angle of attack for an aerofoil

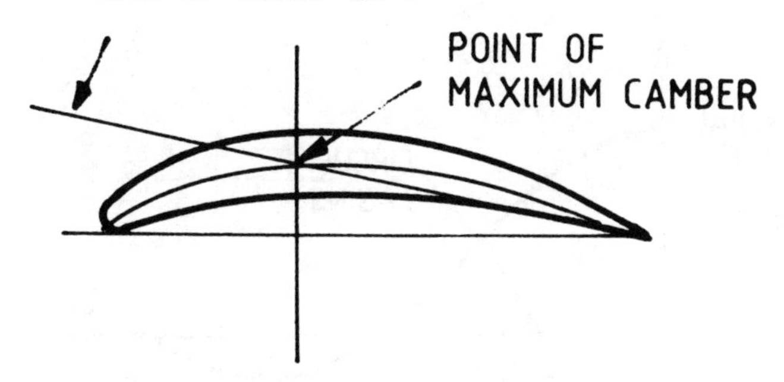

Fig. 2.21(b) A cambered profile

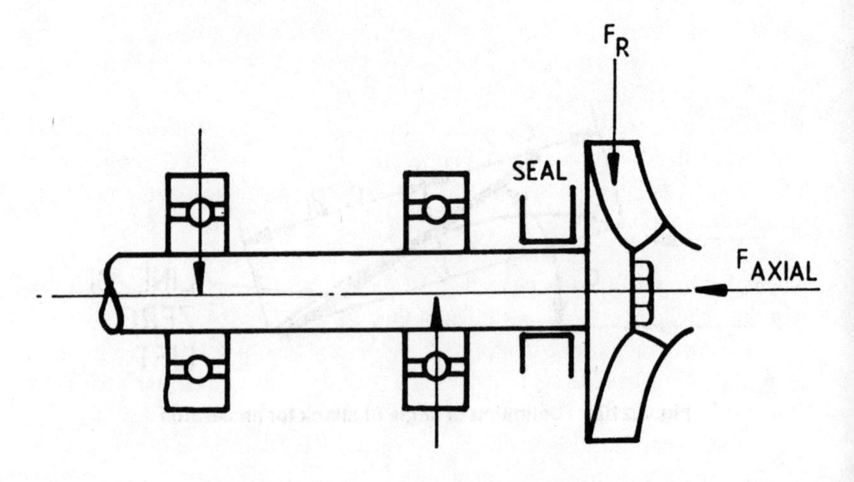

Fig. 2.22 Hydraulic loads on a shaft system

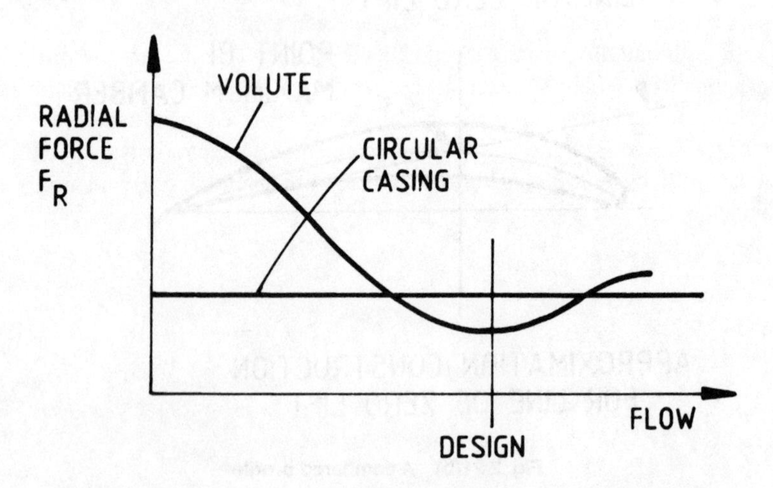

Fig. 2.23 Radial hydraulic load change with flow rate for a volute pump

effects in addition to the weight of the rotating assembly are to be considered: the radial and axial hydraulic loads, as shown in Fig. 2.22. If the flow rate is close to that designed, the pressure around the impeller will be approximately the same. When, however, the flow rate reduces, the pressure distribution changes and results in a large radial load (Fig. 2.23). This load has two components, a steady load, and one which is fluctuating. The result is failure of the shaft if the effects are not considered.

Radial loads can be minimized by providing a double casing or, in cases where the drop in efficiency can be tolerated, a circular casing, which will tend to reduce the load effects, as shown in Fig. 2.24.

In a centrifugal pump with a shroud and backplate provided with a wear ring to limit the flow back to suction there is a pressure effect on both the back plate and the shroud which gives rise to an axial thrust load tending to push the rotating assembly towards the suction pipe (Fig. 2.25). This load, if not balanced out, gives rise to large axial loads which have to be withstood by the thrust bearings. In a single stage suction pump there are two alternative solutions: the balance chamber system (Fig. 2.26) and pump out-vanes provided on the reverse of the impeller back plate. Both methods have their problems: the balance chamber increases the flow back to suction, thus reducing the volumetric efficiency and, hence, the overall efficiency of the pump and the pump out-vanes use more power and thus reduce the overall efficiency in a different way.

In an **axial flow pump** the radial thrust loads are not present since the machine will be in rotational balance, but there is an axial thrust load due to the forces on the blades, shown in Fig. 2.27. This is derived in the same way as the tangential load F_T, but a simple equation will give an approximate value for the axial thrust, by multiplying the annulus area by the pressure rise generated by the impeller

$$F_A = (p_1 - p_2)\,\frac{\pi}{4}\,[D_T^2 - D_H^2]$$

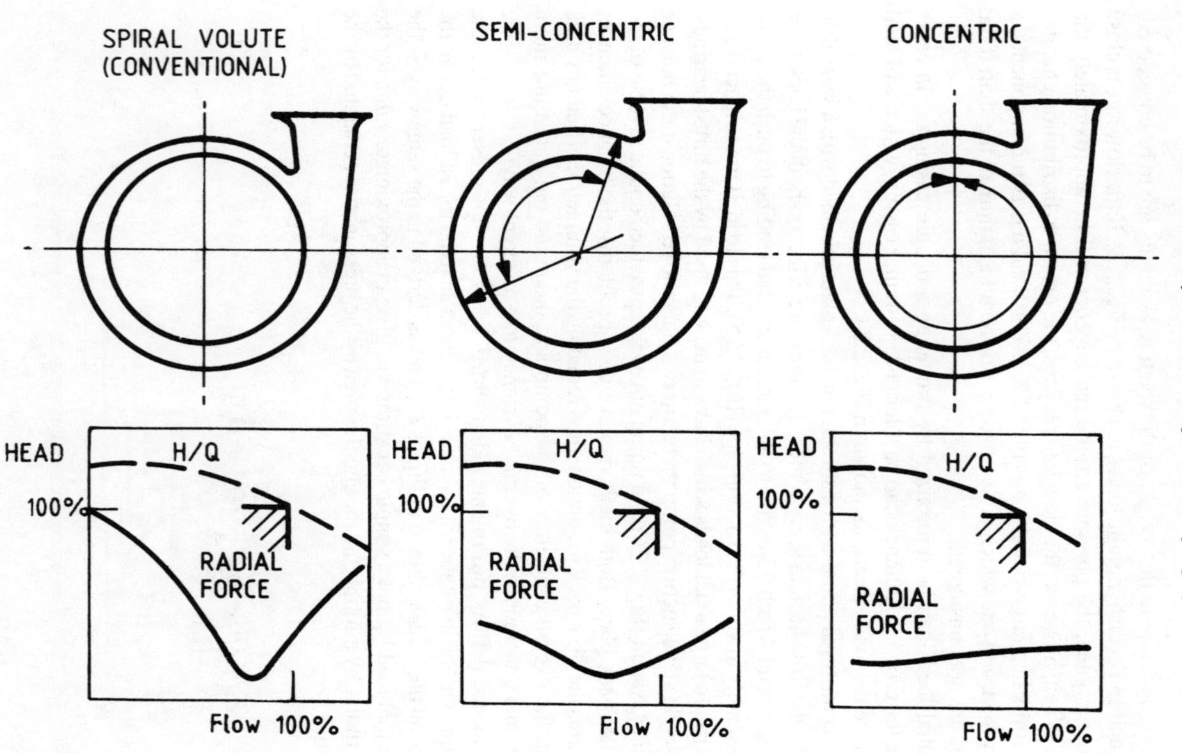

Fig. 2.24 The effect of casing change on radial load

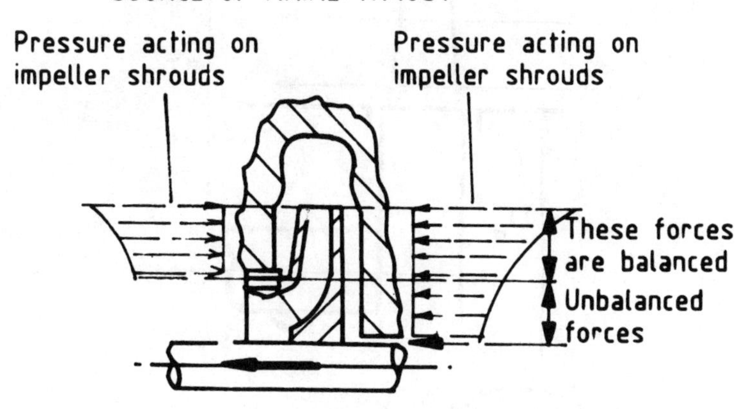

Fig. 2.25 The source of axial thrust in a centrifugal pump

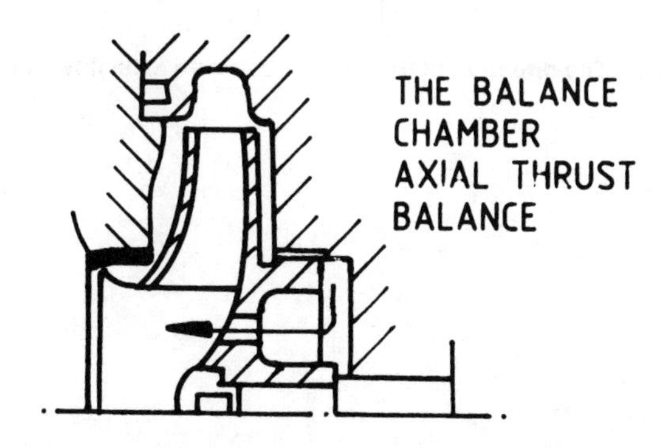

Fig. 2.26 The balance chamber method of balancing axial thrust

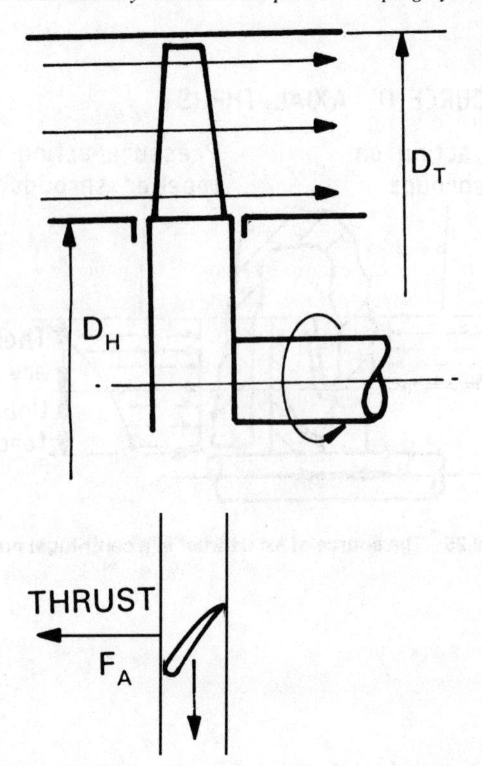

Fig. 2.27 The direction of the axial thrust in an axial flow pump

CHAPTER 3

Positive Displacement Pump Principles

3.1 INTRODUCTION

The main types of positive displacement pump have been described in Chapter 1. The basic principles of these designs will be discussed in this chapter, and the effects of viscosity, gas content, and pulsation will be described. Discussion of the effects of relief valves and pulsation dampers will conclude the chapter.

Figure 3.1 lists the main pumps in the family.

3.2 BASIC PRINCIPLES

The flow rate delivered is related to speed and geometry; for example, a reciprocating, single acting pump shown in Fig. 3.2 would deliver, without leakage

$$Q_o = \omega \times \text{swept volume}$$
(the swept volume = stroke × piston area)

Similar calculations for rotary machines may be performed.

In practice there is always leakage flow through the necessary clearances between the moving and stationary components; thus the actual flow rate will be

$$Q = Q_o - Q_L$$
(Q_L = leakage flow)

The volumetric efficiency is η_V

$$\eta_V = \frac{Q}{Q_o} = 1 - \frac{Q_L}{Q_o}$$

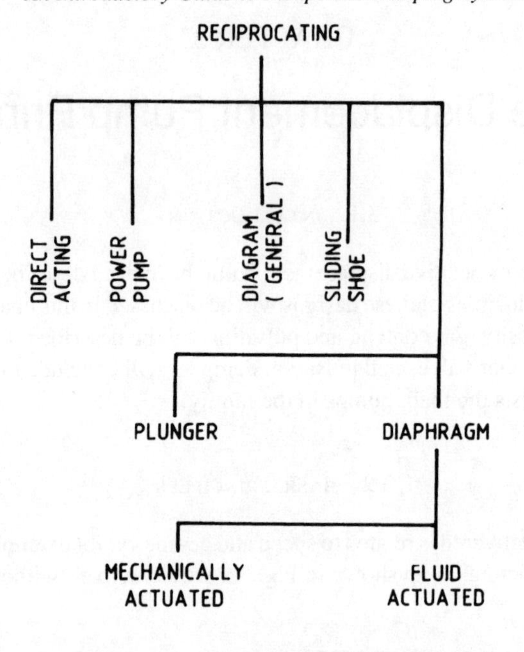

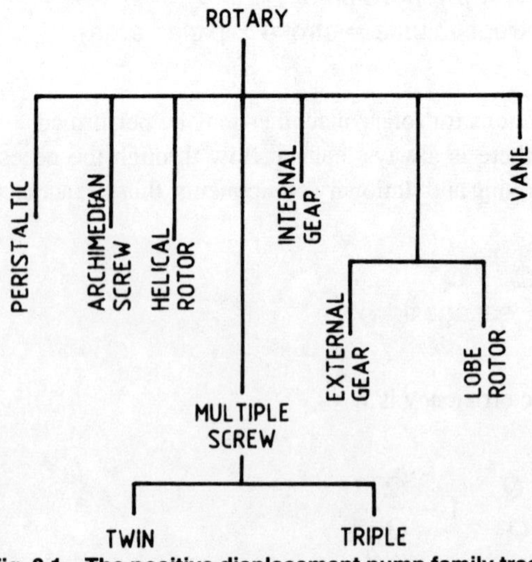

Fig. 3.1 The positive displacement pump family tree

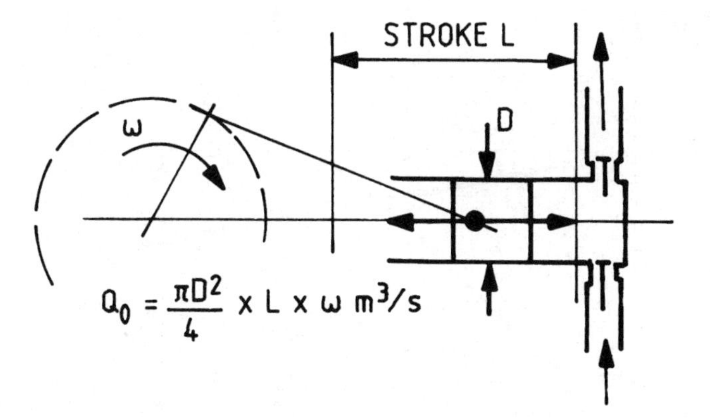

Fig. 3.2 A simple reciprocating pump (with the leading dimensions defined)

The power delivered to the pump is related to work done to the fluid (P_D) and to mechanical and fluid losses (P_F)

Power

$$\eta = \frac{P}{P_D + P_F}$$

The idealised variation of flow rate power and efficiency against pressure rise for constant speed is shown in Fig. 3.3. Table 3.1 gives some typical efficiencies.

Table 3.1 Some typical values of η_V and η

Machine	η_V	η
Precision gear	>98%	>95%
Screw	>98%	75–85%
Vane	85–90%	75–80%
Multi-piston	>98%	>90%

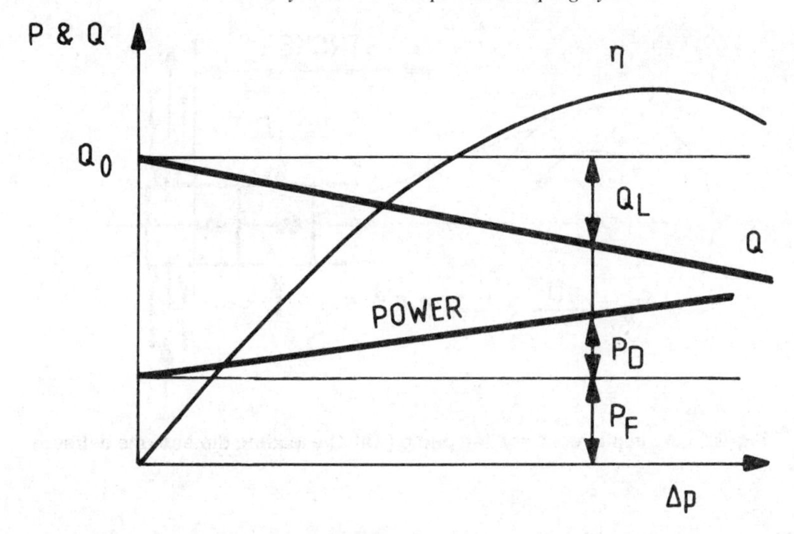

Fig. 3.3 An ideal positive displacement pump ideal characteristic plot

3.3 THE EFFECT ON PERFORMANCE OF PRESSURE, SPEED, AND VISCOSITY

3.3.1 The effect of pressure rise Δp

If the speed and the viscosity is held constant, Fig. 3.4 illustrates the relation between flow rate and Δp. When $Q_L = Q\,\eta_V$ is 50 per cent, this limits effectively the working range that is practicable, and as Δp increases $Q_O = Q_L$, the 'dead end' condition, represents the ultimate of $Q = O$. In practice leakage will bring $Q = O$ at a lower value of Δp.

3.3.2 The effect of speed change

If the viscosity is constant, and Δp is constant, flow rate varies as speed, as sketched in Fig. 3.5. Q_O varies linearly as rotational speed, but Q_L is approximately constant, so the actual flow rate, Q, is the full line. The linear relationship applies, except at high speeds when fluid may not completely fill the pump spaces.

3.3.3 Viscosity change effects

In Fig. 3.6 the effect of viscosity is sketched for a machine at constant stroking speed or rotational speed. In Fig. 3.7 an alternative presentation for a rotary machine is given, for constant Δp and speed. Q_O will not change with μ if viscosity is high, (leakage flow will be laminar), so Q_L is a straight line from the point $Q = Q_L$ for higher viscosity. If the liquid is of lower viscosity leakage flow is turbulent, and Q_L varies as shown. One effect is that of temperature rise and the dotted line indicates the change in Q that could occur.

3.4 THE EFFECT OF GAS CONTENT

Liquids always contain dissolved gases, which as pressure falls emerge out of solution as bubbles which thus occupy space that should be full of liquid. The pump flow rate is thus adversely affected, and the reduction is obtained using Fig. 3.8. Additionally if the suction system is poor bubbles are entrained, and this effect is indicated by Fig. 3.9.

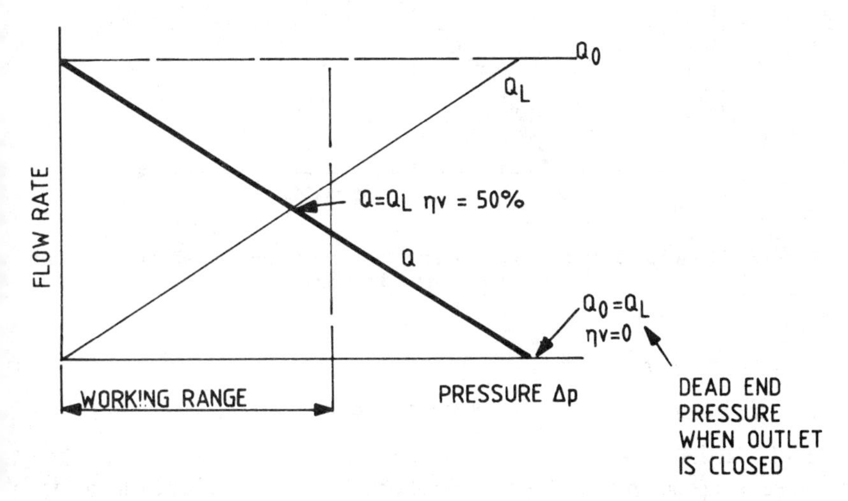

Fig. 3.4 The effect of pressure change on for a rotary pump with the speed and viscosity held constant

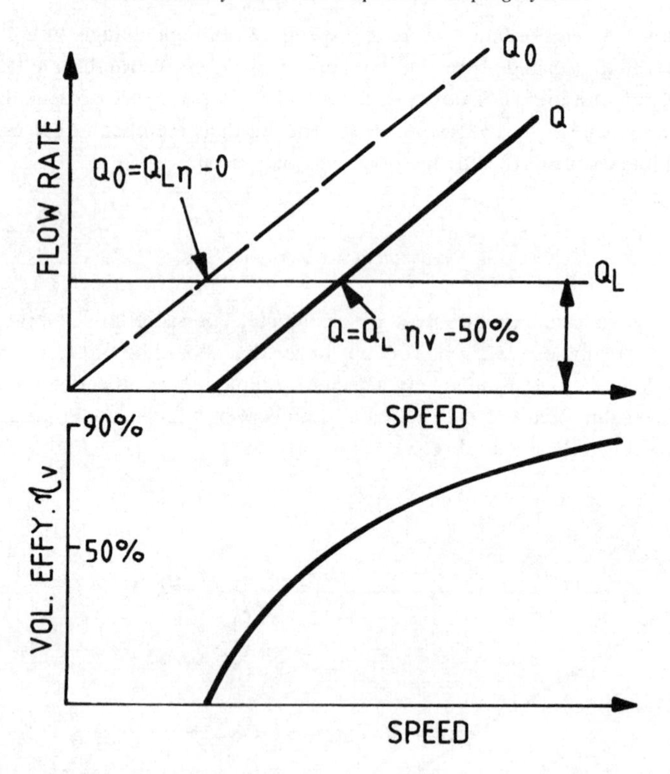

Fig. 3.5 The effect of speed change on a rotary pump flow rate with viscosity and pressure rise held constant

3.5 PUMP PROTECTION

3.5.1 Relief valves

Some pumps are provided with internal or integral relief valves, which cope with intermittent operation against closed valve. If operations give high differentials often, or integral valves are not fitted, a relief valve connected to the suction vessel is required to protect the pump. Relief valves should reseat, and a desirable valve characteristic is shown in Fig. 3.10 in this figure: 'a–b' is the pump characteristic without the valve; point 'c' is the set pressure; point 'd'

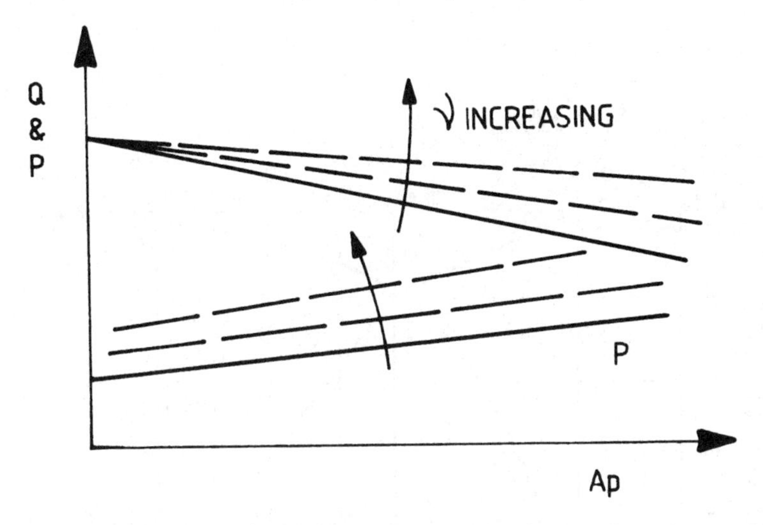

Fig. 3.6 The general effect of viscosity change on the normal characteristics for a positive displacement machine

denotes maximum pressure differential; point 'f' denotes the instant when closing-response accelerates; point 'g' is the reseating pressure.

The over-pressure (point 'd') required to open the relief valve completely will depend upon the product being handled and the geometry and resistance of the return pipe. Typical accumulation pressure may well be 25 per cent above the set pressure at point 'c'. At point 'd' all the discharge from the pump (represented by the ordinate 'd–e') will be flowing through the relief valve and back to the suction branch or supply tank. The over pressure Δp_d will be broken down across the relief valve and through the flow resistance offered by the line returning the product to the suction side.

Should the pressure in the delivery pipe line drop to point '*h*' in Fig. 3.10 after full relief valve opening, then the pump is able to start to supply a small quantity 'h–f' to the line. The major part of the product output from the pump will, however, continue to be returned to suction via the open relief valve. Further reduction in discharge pressure will enable the valve to start to return to its seat, and this reseating will be complete at point 'g'. The reseating pressure will be

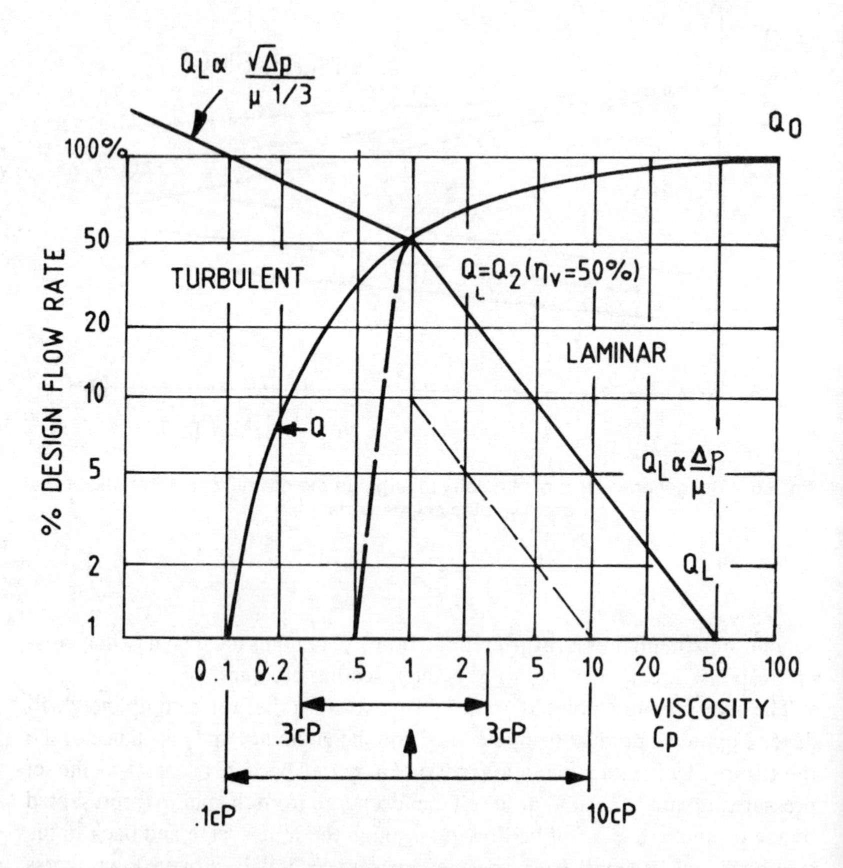

Fig. 3.7 The effect of viscosity change for a rotary pump with the pressure rise and viscosity held constant

less than the set pressure by an amount called the 'blow-down' pressure which again can be as much as 10 or 15 per cent of set pressure. Accumulation pressure and blow-down pressure need to be kept as small as possible, since they together make the Δp range 'g–e' unavailable for pump operation.

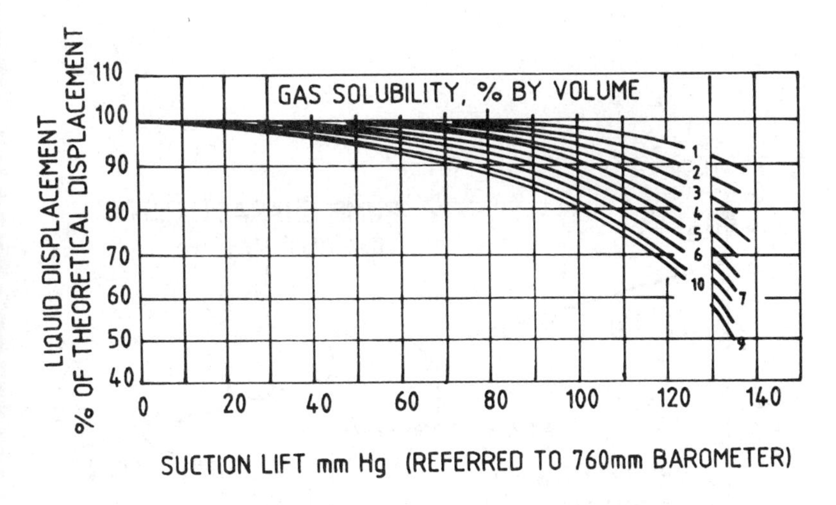

Fig. 3.8 The effect of dissolved gas on the volumetric efficiency of a rotary pump

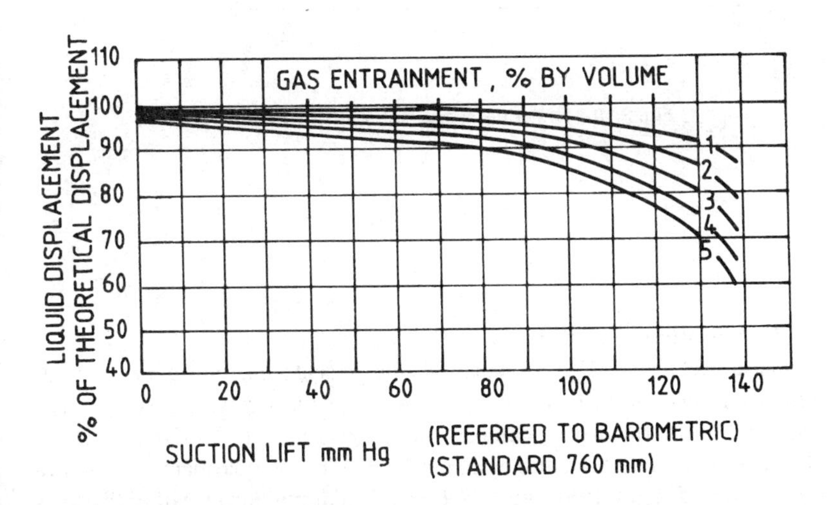

Fig. 3.9 The effect of gas entrainment on the volumetric efficiency of a rotary pump

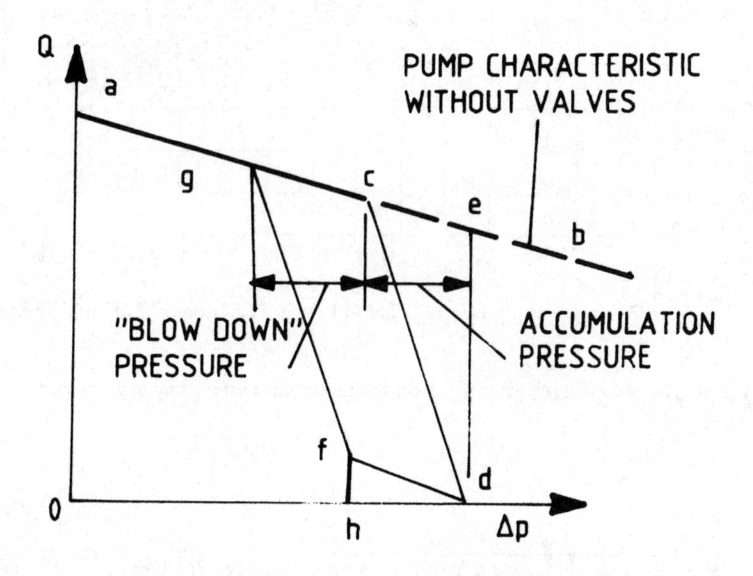

Fig. 3.10 The general characteristic of a relief valve

3.5.2 Pulsation control

Reciprocating pumps (Fig. 3.11) in particular present a fluctuating pressure pattern in both suction and delivery, and even though the use of multiple cylinders does reduce the fluctuating component, dampers need to be fitted, to either attenuate or absorb the fluctuations. Figure 3.12 illustrates the main types, with a comparison of their effectiveness. The excellent discussion by Miller (**9**) gives details of proportions and pressure limits.

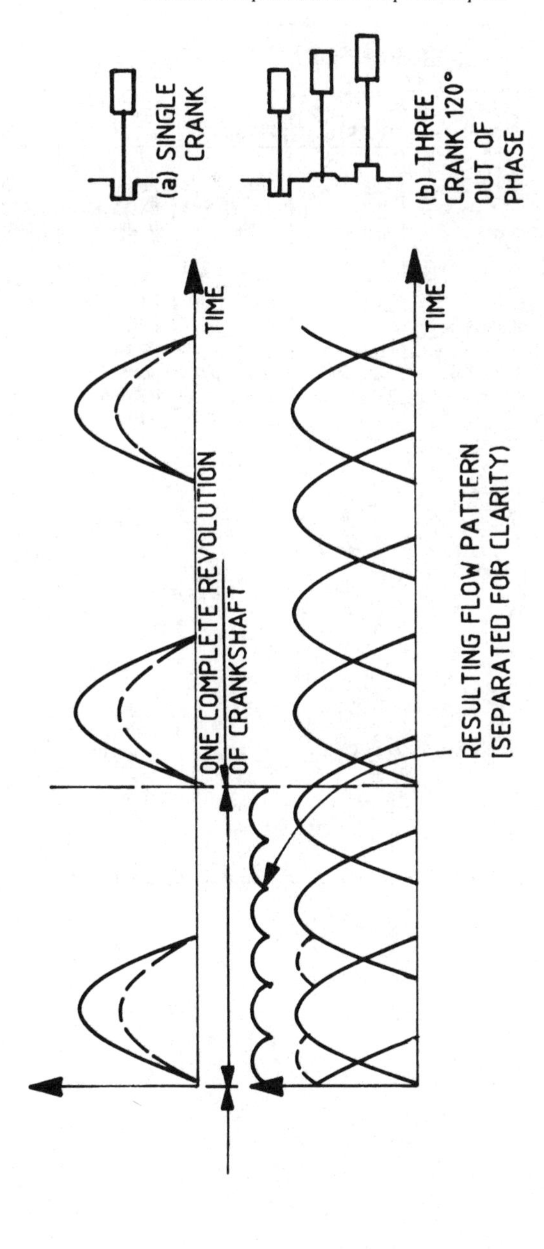

Fig. 3.11 The effects of pulsation on the delivery pressure for a single cylinder reciprocating pump and for a triplex design

RELATIVE DAMPENER PERFORMANCE

	DIVERTER	APPENDAGE	GAS-OVER LIQUID	ACOUSTIC
ATTENUATION:				
HIGH FREQUENCY	FAIR	POOR	POOR	EXCELLENT
LOW FREQUENCY	EXCELLLENT	EXCELLENT	EXCELLENT	POOR
BACK PRESSURE ON PUMP	SLIGHT	NONE	NONE	HIGH
EFFECTIVENESS AT;				
VARIOUS PUMP rpm	EXCELLENT	EXCELLENT	EXCELLENT	POOR
VARIOUS PUMP gpm	EXCELLENT	EXCELLENT	EXCELLENT	POOR
VARIOUS PUMP PRESSURES			EXCELLENT	FAIR
INITIAL COST	HIGH	LOW	MEDIUM	HIGH
REPAIR COST	MODERATE	MODERATE	NONE	NONE
MAINTENANCE COST +	LOW	LOW	MEDIUM	NONE
SLURRY CAPACITY	POOR	EXCELLENT	GOOD	POOR

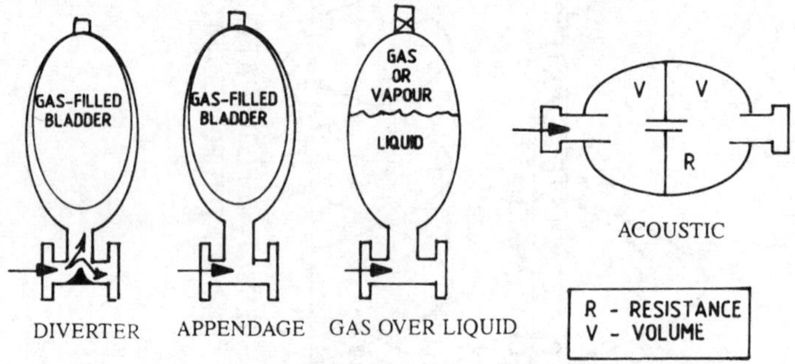

Fig. 3.12 The relative performance of four damper designs from Miller (9)

CHAPTER 4

Pumping System Losses and their Estimation

4.1 INTRODUCTION

In any pumping system resistance to flow is due to friction, and to those losses due to flow disturbance caused by features such as bends, valves, junctions, and other essential pipework devices, and also to the need to lift liquids from one level to another (static lift).

As Osbourne Reynolds demonstrated, friction losses vary as flow rate squared, so that a general equation for dynamic loss of energy is

$$(gH) = K\,V^2/2$$

Here V is found by dividing the volumetric flow rate by the area of flow, and the constant K is a function of the fluid flowing, and the geometry of the duct through which flow is taking place. The losses due to friction and to other features will now be discussed using this form of equation.

4.2 LOSSES DUE TO FRICTION

An equation found in many textbooks is the D'Arcy–Weisbach relation

$$(gH) = f\frac{L}{D}\frac{V^2}{2}$$

where L is the length of the pipe being used, D its diameter, V the mean velocity of flow, and f the friction factor.

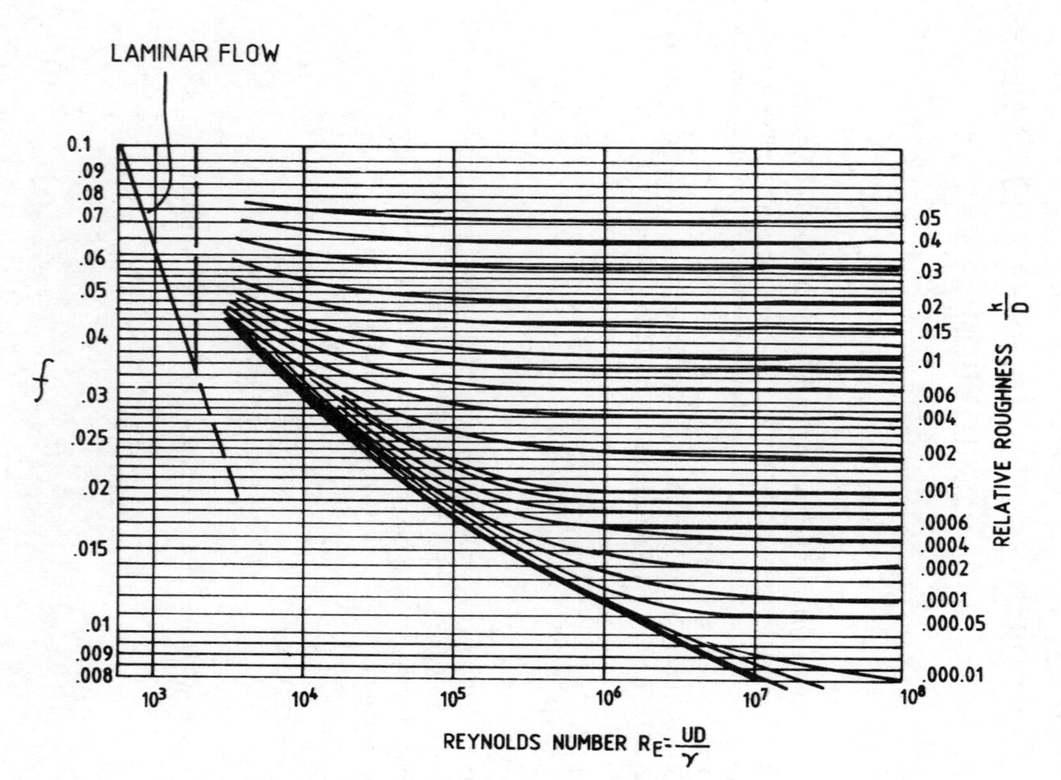

Fig. 4.1 The Moody diagram (after the plot presented by Miller (10)

The friction factor *f* is related to the fluid properties, and to the surface roughness of the pipe line. The Moody Diagram (Fig. 4.1) is a plot of the friction factor against Reynolds number and relative roughness. The relative roughness ratio uses the average surface roughness, typical values of which are shown in Table 4.1.

Table 4.1 Some typical hydraulic roughness values

Type of pipe	*Hydraulic roughness,* k *(mm)*
Cast iron	0.203
Galvanised steel	0.152
Uncoated steel	0.051
Coated steel	0.076
Drawn brass, copper, aluminium, glass, plastic	Smooth
All welded pipe	0.076
Flexible pipe	If smooth rubber or plastic as steel pipe. Head loss 30% higher when curved. If corrugated refer to maker

The dimension D for a circular pipe is, of course, the diameter of the pipe; for a non-circular duct the 'equivalent diameter,' D_e, is used, defined as

$D_e = 4 \times$ area of flow/wetted perimeter

(For a circular pipe running full

$$D_e = 4 \times \pi \frac{D^2}{4} / \pi D = D)$$

As an example of the way the loss due to friction can be determined, consider the loss in a pipe line 500m long and 50mm diameter through which water at normal temperature is flowing at the rate of $0.01 m^3 s^{-1}$.

Mean velocity $= 5.09$ m s^{-1}

The kinematic viscosity of water at normal temperature is $1 \times 10^{-6} m^2 s^{-1}$, so the Reynolds Number R_e becomes 2.545×10^5.

Since the pipe is considered to be uncoated steel the roughness is 0.051mm from Table 4.1, the relative roughness is 0.00102.

From Fig. 4.1 $f = 0.020$.

Loss to friction is thus

$$0.020 \times \frac{500 \times 5.09^2}{0.050 \times 2} = 2591 \text{ J kg}^{-1}$$

The loss expressed (as is common practice) in metres of water is 264.

4.3 LOSSES DUE TO FLOW THROUGH BENDS, VALVES, AND OTHER FEATURES

4.3.1 The *K* factor method

One method of allowing for losses due to fittings is to use the formula as given in section 4.1, and to assign values to the factor *K* so that individual losses can be estimated. The text by Miller presents a comprehensive study of these losses and also a series of plots of correlations of the *K* factors.

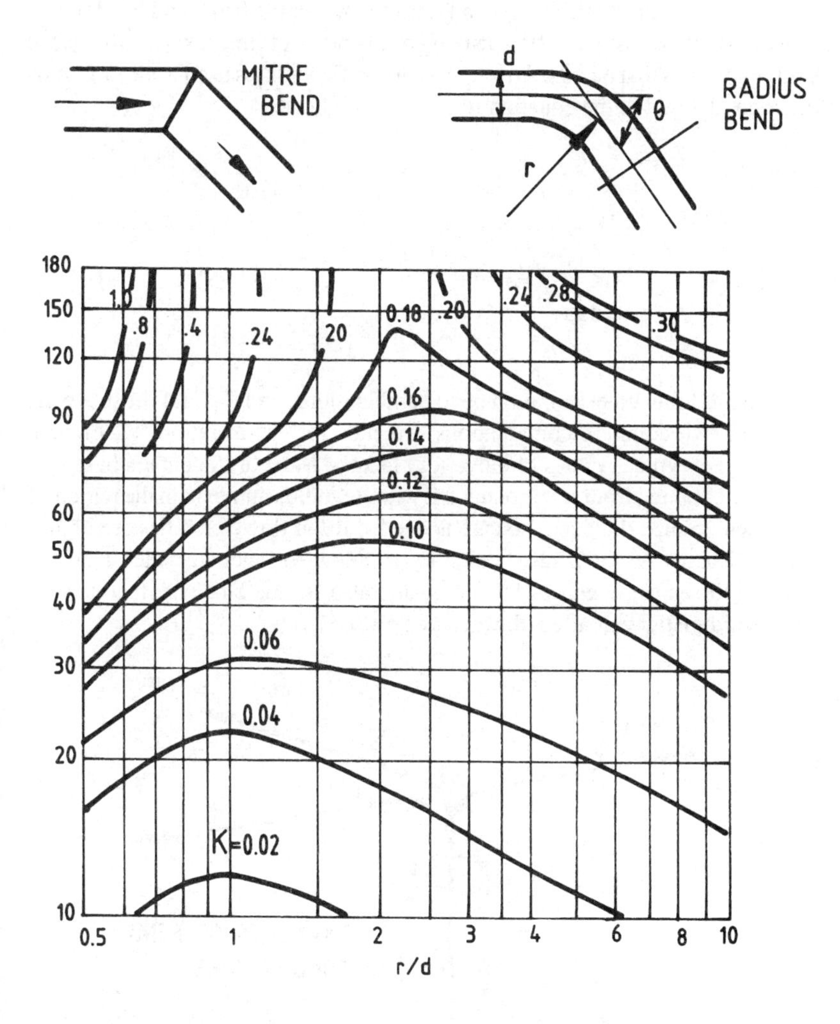

Fig. 4.2 Bend loss factors (after the plot presented by Miller (10)**)**

Figure 4.2 gives data for K factors for bends presented in terms of the bend-to-radius ratio and the angle of the bend. This is for a single bend, and bends come in sets so that the disturbed flow leaving one bend meets the next and affects the consequent loss that occurs. An interaction coefficient K_p is used to allow for this problem, defined in the equation relating to two bends

$$(gH) = K_p \left(K\frac{V^2}{2} + K\frac{V^2}{2}\right)$$

Bend 1 loss Bend 2 loss

Figure 4.3 illustrates common bend combinations, and Fig. 4.4 illustrates K_p value for two common combinations of bends for an r/d ratio of 2. Figures 4.5–4.7 give typical values of valve loss factors. Most of the data is based on evidence obtained from tests conducted with Reynolds numbers in the region of 10^5 based on pipe diameter, so care is needed if the flow velocities are higher than typical values. It is also necessary to allow for ageing, which tends to increase losses as roughness increases, and also for the blockage that occurs, when solid matter can affect the area of flow.

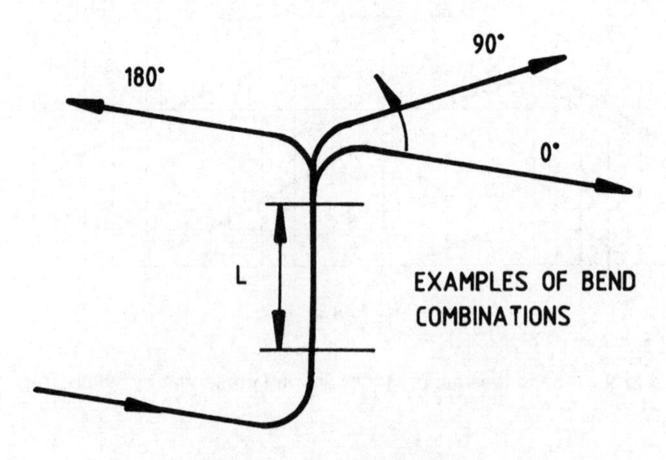

Fig. 4.3 Bend combinations

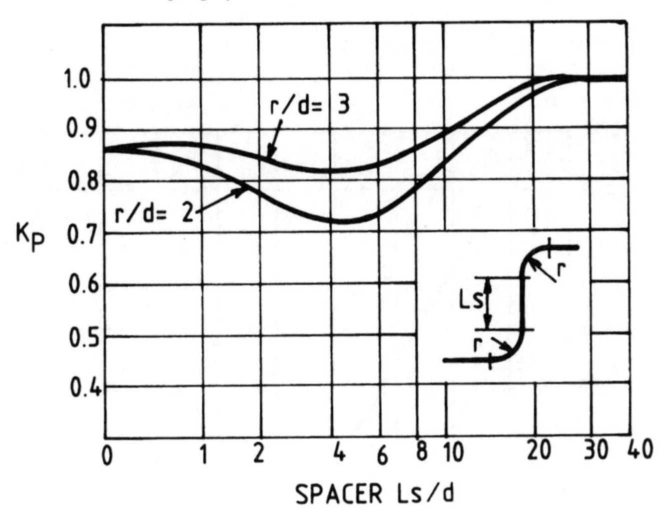

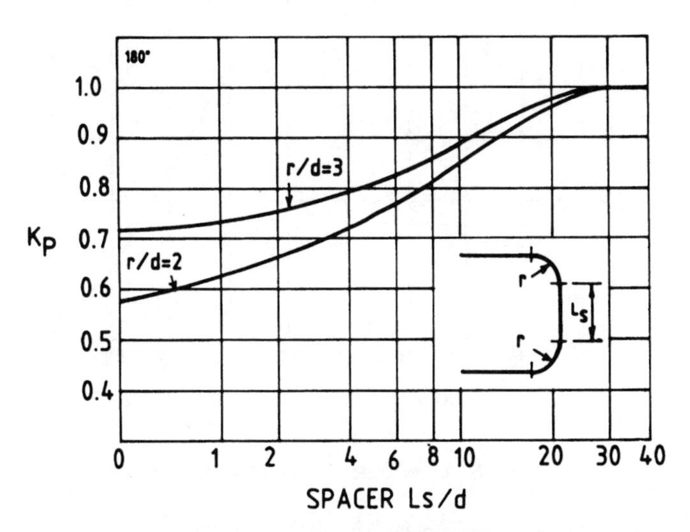

Fig. 4.4 Bend loss correction factors, K_p, for bend combinations
Top The 0 degree combination
Bottom The 180 degree combination

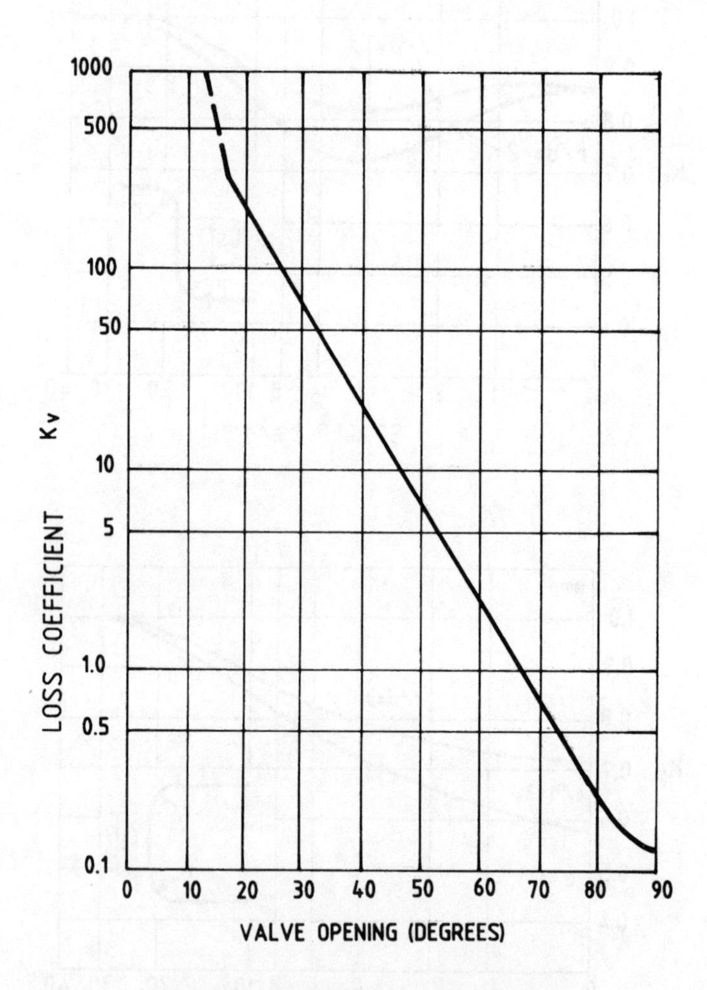

Fig. 4.5 Loss factor for a butterfly valve

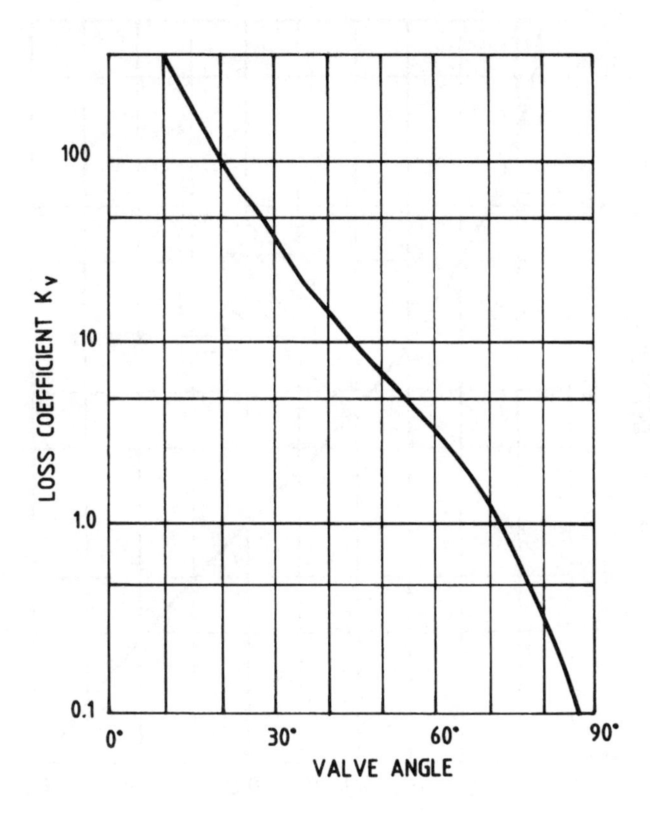

Fig. 4.6 Loss factor for a spherical valve

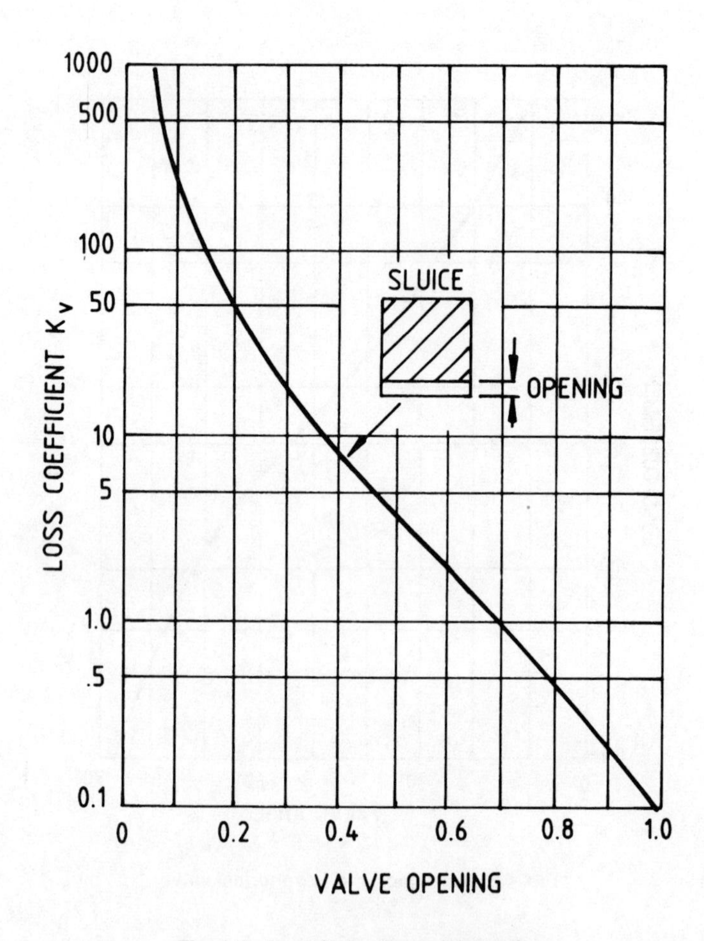

Fig. 4.7 Loss factor for a sluice valve

Reference must be made to reference manuals such as that by Miller (**10**) for the losses due to other features.

4.3.2 The equivalent length method

As the name suggests, the loss a feature causes is replaced by the loss due to a length of straight pipe-line which gives the same energy loss. Table 4.2 gives some values of the equivalent lengths for a number of features, L_E being the equivalent length in metres of straight pipe.

Table 4.2 Some typical equivalent length L_E values expressed in pipe diameters

Feature	L_E
Gate valve (fully open)	13
(half shut)	160
Ball valve (fully open)	340
90° bend	30
90° long radius bend	20
Butterfly valve (fully open)	40

4.4 TYPICAL SYSTEM CHARACTERISTICS

If the simple system shown in Fig. 4.8 is considered, the pump must overcome the static lift before any flow can be generated. Once flow starts the flow or dynamic resistance, which varies as the square of the flow rate, must also be supplied. Typical system characteristics are shown in Fig. 4.8. Also shown in the figure are the possible alternative system resistances due to short and long pipelines.

4.5 SYSTEM NET POSITIVE SUCTION HEAD (NPSH) OR NET POSITIVE SUCTION ENERGY (NPSE)

For many years pump engineers have used the term Net Positive Suction Head (NPSH) to give the margin between the local pressure head and the vapour

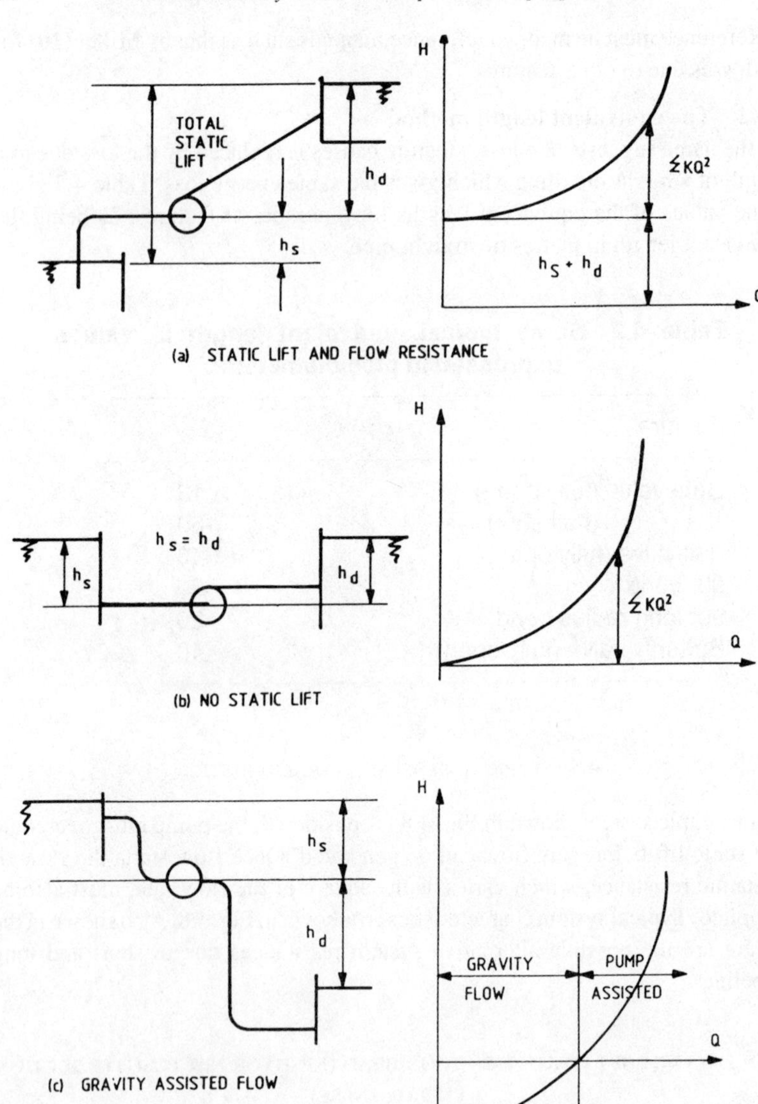

Fig. 4.8 Typical systems and their characteristics

pressure head at the point in the flow path considered as an index of the possibility of cavitation occurring. With the introduction of the concept of specific energy, the term used is the Net Positive Suction Energy available or $NPSE_A$, this is defined as

$$NPSE_A = \text{Total Specific Energy at Suction Flange} - \text{Vapour Pressure Energy}$$

The total specific energy or NPSHA is calculated for a suction system by using the approach illustrated in Fig. 4.9

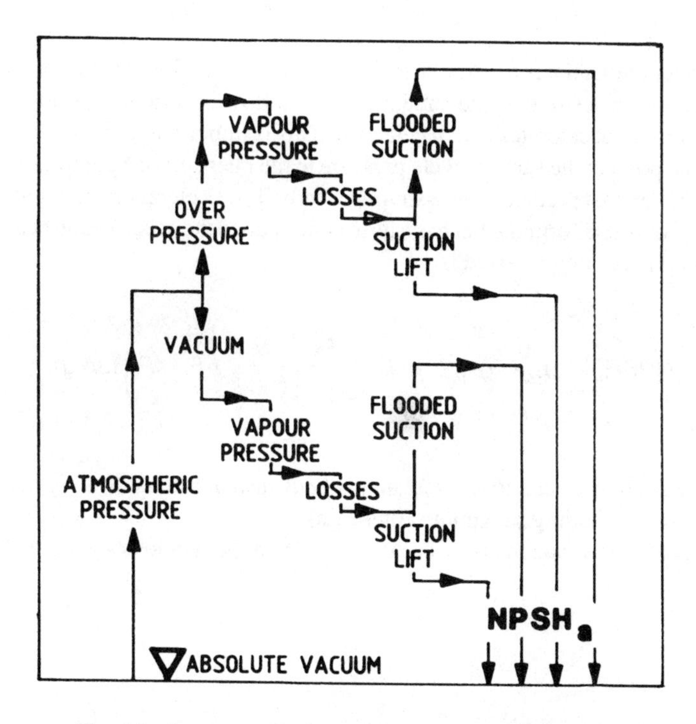

Fig. 4.9 An approach to the calculation of NPSH available

In a manual edited by Davidson (**3**) the $NPSE_A$ is given by the formula

$$NPSE_A = gh_{Si} - gH_{Ri} + \frac{100}{\rho}\left(\frac{B}{1000} + P_i - P_v\right) \text{ (J/kg)}$$

where

h_{Si} = static liquid head *over* pump inlet (m)
B = barometric pressure (mbar)
H_{Ri} = frictional losses in suction system (calculated at *max* flow)
P_i = gas pressure on free surface (bar *gauge*)
P_v = vapour press of liquid at *max* operating temp
ρ = density (kg/l)

This equation relates to rotodynamic machines where flow can be considered continuous. In positive displacement machines flow is periodic, so flow in the piping system accelerates and decelerates. This is important in the suction line, for at one point in the suction cycle pressure in the inlet port of a pump can fall to a level below that predicted by assuming steady flow. Reference (**3**), mentioned above, proposes a formula that has a term added called the acceleration head, H_{ai}. A general equation proposed is

$$NPSE_A = gh_{Si} - gH_{ri} - gH_{ai} + \frac{100}{\rho}\left(\frac{B}{1000} + P_i - P_v\right) \text{ (J/kg)}$$

H_{ri} is the frictional head loss calculated at *peak* instantaneous flow (rather than mean flow) (including maximum filter loss).

The peak instantaneous flow is calculated from the relations shown in Table 4.3.

Table 4.3

For a reciprocating pump	*No. of pistons*	*Peak flow/ Mean flow*
	3	1.115
	5	1.035
	7	1.018
For metering pumps		
	Quintuplex	1.03
	Triplex	1.1
	Duplex	1.6
	Simplex	3.2
For rotary pumps		
Any type with a damper in the line:		
Screw pumps and helical gear pumps		1.0
Mono pumps		1.03
Vane pumps		1.05
Two-lobe pumps		1.25
Peristaltic – planar cam		2.0
– roller type		1.5

The acceleration head H_{ai} is used as follows.

For reciprocators (without damper fitted)

$$gH_{ai} = \frac{6867.N.Q}{Z} \Sigma \frac{l}{d^2} \text{ (m)}$$

where

Z = Number of pistons
Q = Flow rate litre/s
N = Crank rotational speed rev/s
l = Length of line (m)
d = Line diameter (mm)

For rotary pumps (without damper fitted)
$H_{ai} = \delta P/p$

$$\delta P = \frac{K_2 \rho C Q}{d^2} \text{ (bar)}$$

where
K_2 = 2 for 3 lobe pumps, 5 for 2 lobe pumps;
5 for a mono pump, 1 for spur gear;
5 for a vane pump, 20 for a planar cam peristaltic;
10 roller type peristaltic, 0 for helical and helical gear pumps.

For reciprocating metering pumps,

$$gH_{ai} = \frac{58.9 \delta P_i}{p}$$

$$\delta P_i = \frac{K \rho CQ}{d^2}$$

where
K = 40 for simplex,
20 for a duplex,
3 for triplex,
1 for quintuplex.
and H_{ai} is as used in the following equations.

For simplex and duplex machines

$$\text{NPSE}_A = \text{Basic NPSE}_A - (gH_{ai}^2 + gH_{ri}^2) - 1$$

For triplex and quintuplex machines

$$NPSE_A = \text{Basic } NPSE_A - (gH_{ai} + gH_{Ri}) - 1$$

4.6 THE EFFECT OF FLUID PROPERTIES

The loss equation and the data for valve and bend loss are all based on empirical information obtained for Newtonian fluids which all obey the rule

$$\tau = \mu \frac{\delta V}{\delta y}$$

where

τ = shear stress (N/m^2)

μ = coefficient of dynamic viscosity (N s/m^3)

$\frac{\delta V}{\delta y}$ = velocity gradient (s^{-1})

It is fortunate that many engineering fluids are Newtonian, so that whatever the velocity gradient, μ is constant. A large number of chemicals, however, are non-Newtonian, and Fig. 4.10 illustrates their properties.

Bingham plastics are found in solids transport systems, and their characteristic is that they need some shear, τ_O, to be overcome before they will flow; after this initial shearing action they behave as Newtonian fluids.

Pseudoplastics follow a law

$$\tau = K\left(\frac{\delta V}{\alpha y}\right)^n \text{ where } n > 1$$

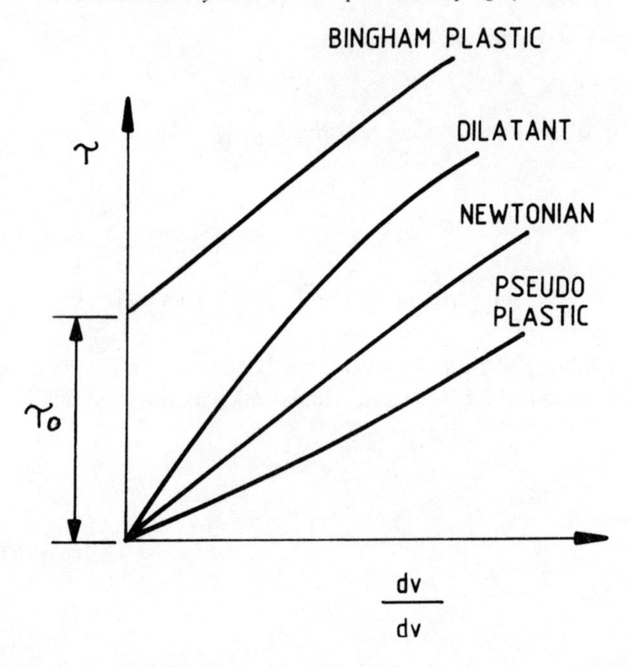

Fig. 4.10 A generalised chart of the types of fluid viscous resistance

and has an apparent viscosity

$$\mu_A = K\left(\frac{\delta V}{\delta y}\right)^{n-1}$$

Examples are melted polymers and dye pastes. Dilatants follow the law

$$\mu_A = K\left(\frac{\delta V}{\delta y}\right)^{n} \text{ where } n < 1$$

An example is starch solution.

A further complication with non-Newtonian liquids is their time-dependent behaviour. For example, thixotropic fluids like printing ink and non-drip paints exhibit a reducing τ with time at a constant shear rate or velocity gradient, and μ_A increases again as the fluid velocity reduces.

Some slurries, like gypsum suspensions, are rheopectic, and thicken rapidly if shaken or disturbed.

Texts on rheology should be consulted for information on the methods used to determine μ_A and K.

CHAPTER 5

Cavitation Effects on Pumps

5.1 INTRODUCTION

Cavitation is a term loosely applied to the formation of gas bubbles in a flowing liquid. Usually such bubbles will form when the local pressure falls to about vapour pressure. All liquids will, according to Henry's law, absorb air through a free surface, the amount being a function of the pressure on the free surface. In addition to this effect many process liquids are pumped at temperatures which are high, like condensate, for example, and due to the process carry bubbles in suspension. These fluids will cavitate easily in such areas as the suction zone of pumps where, due to the pumping action, pressure is low. Bubble formation and collapse has been shown by calculation and observation to occur in micro-seconds. Since bubbles tend to form and collapse in impeller passages, this has a considerable dynamic effect on the column of flowing liquid, giving rise to hydraulic performance reduction. Furthermore, bubble collapse gives rise to

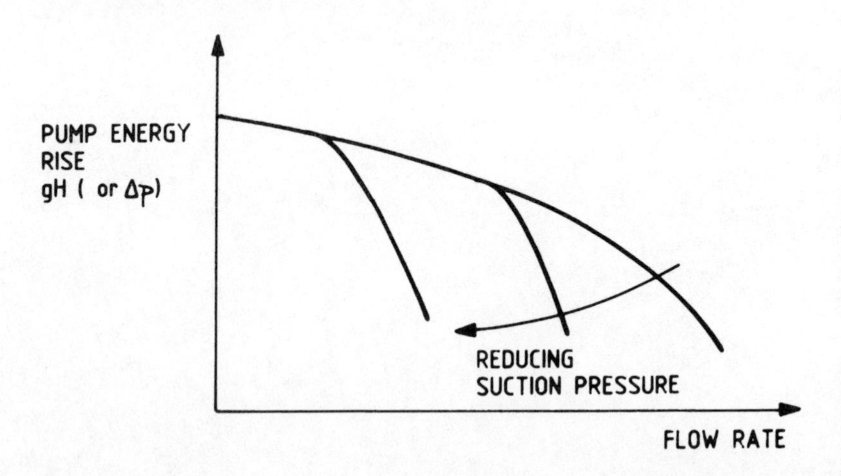

Fig. 5.1 The effect on a centrifugal pump characteristic of reducing the suction pressure

surface damage. Both these effects are discussed in this Chapter, and the criteria used to predict the possibility of cavitation are described. Methods of monitoring pump 'health' are discussed, and a brief discussion of design approaches minimizing cavitation concludes the chapter.

5.2 CAVITATION EFFECTS

5.2.1 Hydraulic effects

The presence of bubbles in impeller passages effectively acts as a fluid valve, and as suction pressure falls the centrifugal pump characteristic flow range is progressively limited, as shown in Fig. 5.1.

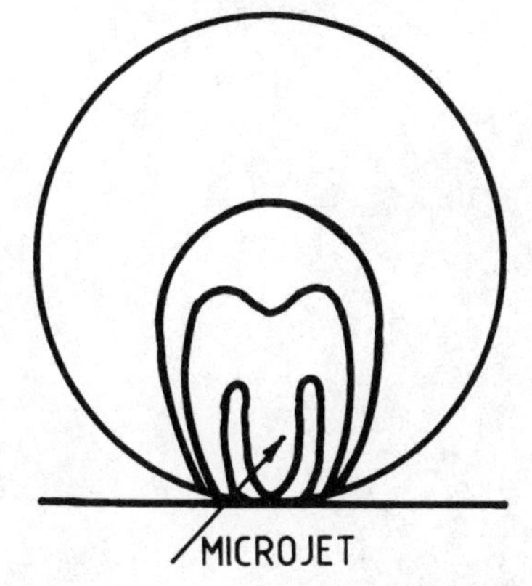

COLLAPSE OF BUBBLE SHOWING FORMATION OF MICROJET (TOP) AND DETAIL OF MICROJET IMPACT (BOTTOM)

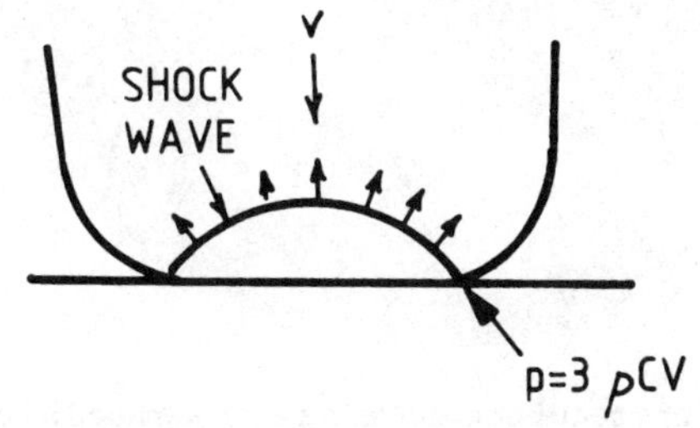

Fig. 5.2 The microjet principle underlying bubble collapse (after Lush (11))

Positive displacement machines exhibit a decrease in volumetric efficiency, as spaces that should be full of liquid are bubbles instead.

Rotodynamic machines exhibit an increased noise spectrum due to the dynamic effects, as will be discussed later in section 5.4.

5.2.2 Surface damage

Surface damage is considered to be due to bubble collapse as shown in Fig. 5.2. The jet of liquid is considered to hit the surface at the local speed of sound,

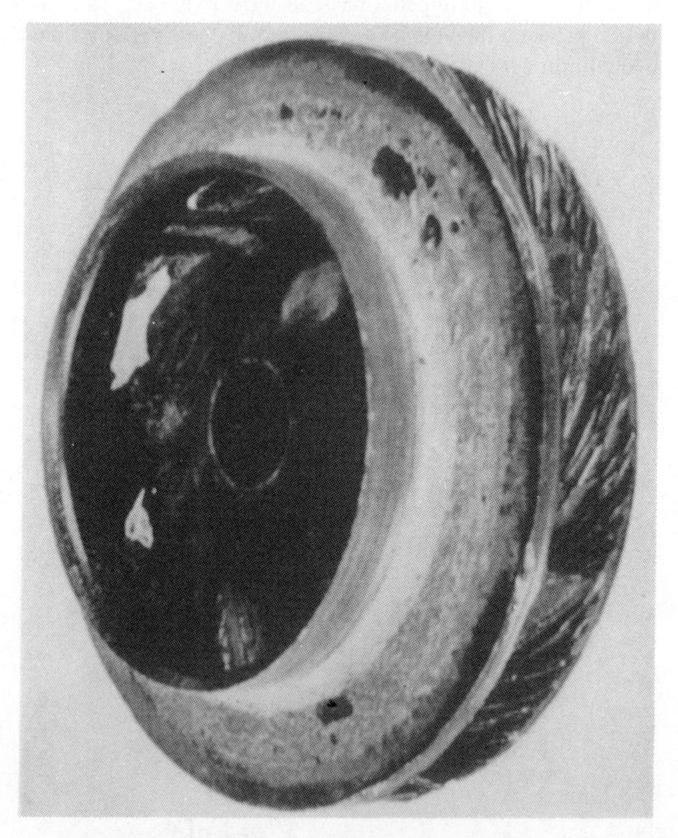

Fig. 5.3 A photograph of a centrifugal pump impeller damaged by cavitation (taken from the James Clayton lecture given by Knap (12)**)**

creating a local surface pressure which is very high, and which can be higher than the ultimate strength of the material. As a result surface break-up followed by erosion and corrosion leads to the pitted surface shown in Fig. 5.3. Table 5.1 indicates how a calculated surface stress compares with several material properties. Figure 5.4 indicates areas where surface damage can occur.

For further information the books and articles listed in the References and the Bibliography should be consulted.

Table 5.1 A comparison of calculated surface stresses with the material properties of a number of materials used in pumps

Microjet impact stresses (after Lush) (11)

Ambient pressure (MPa)	*Collapse velocity (m/s)*	*Maximum pressure (MPa)*
0.1	128	560
1.0	405	1800

Typical strengths of some engineering materials

Material	*Tensile strength (MPa)*	*Hardness (MPa)*
Aluminium	90	270
Cast iron	220	2000
Gunmetal	290	960
HT brass	400	1200
Mild steel	460	1300
Aluminium bronze	470	1600
Stainless steel	790	2500
Monel	830	2900
Titanium	1000	3300

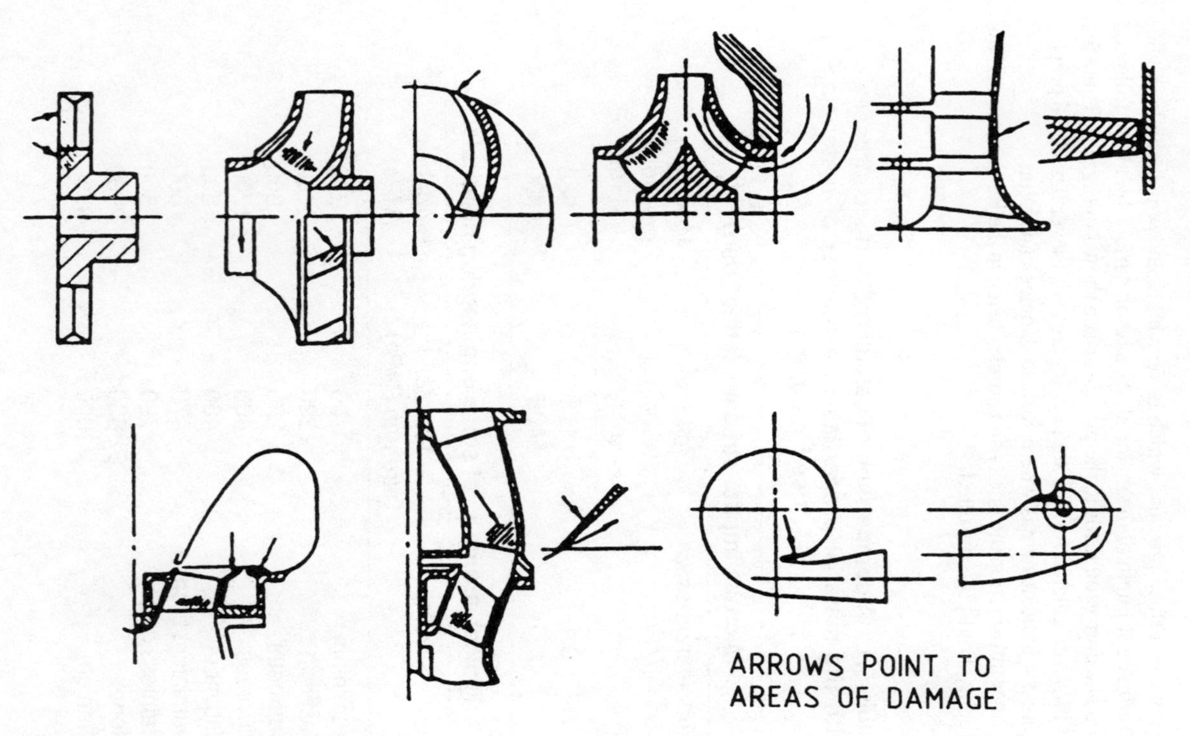

Fig. 5.4 A sketch showing the areas in rotodynamic pumps in which cavitation damage can occur

5.3 CRITERIA USED TO PREDICT THE HYDRAULIC EFFECTS

5.3.1 For rotodynamic machines

The conventional approach is to determine the NPSE that the machine can generate (called $NPSE_{REQUIRED}$). This varies with flow rate. If this is compared with the $NPSE_{AVAILABLE}$ discussed in Chapter 4, and plotted on a base of flow rate as in Fig. 5.5, the pump is not badly affected by cavitation if the flow rate is to the left of the point where the $NPSE_A$ and $NPSE_R$ cross, called the critical flow rate. A further method of presentation required by power test codes and such industry standards as API 610 is to test the pump at constant speed and delivering a constant flow rate (this will be at least design flow rate) with a reducing suction pressure ($NPSE_A$). If Fig. 5.6 is considered, the point at which the machine energy rise, gH falls by x per cent (usually 3 per cent) is regarded as the point at which the machine behaviour becomes critical.

A much used empirical method of presenting this data is to use the critical $NPSE_R$, find the Thoma cavitation parameter, and plot this against characteristic number, as in Fig. 5.7.

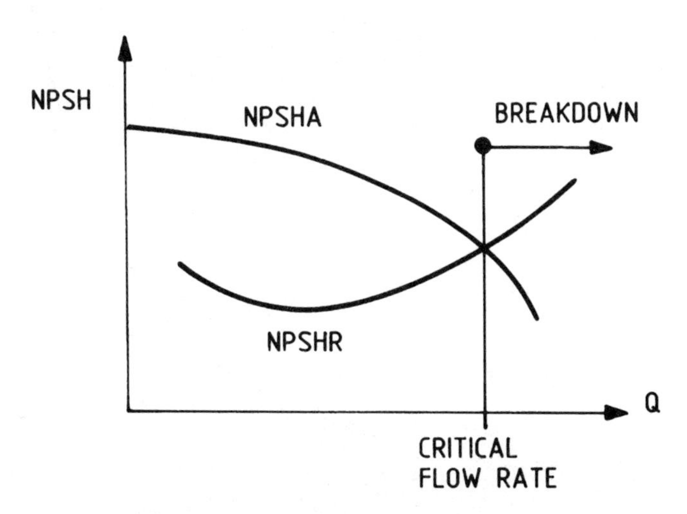

Fig. 5.5 $NPSH_A$ and $NPSH_R$ plotted against flow rate to illustrate the critical flow rate

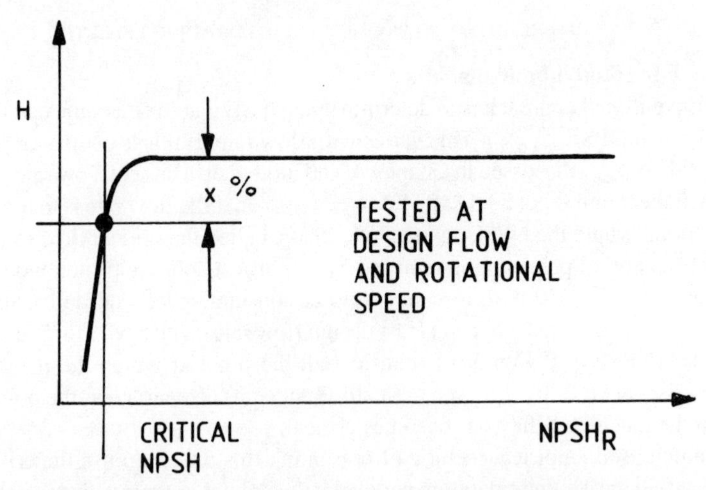

Fig. 5.6 A plot obtained in the usual manner of head variation with $NPSH_R$ when the pump is tested at the design flow rate when driven at the designed rotational speed

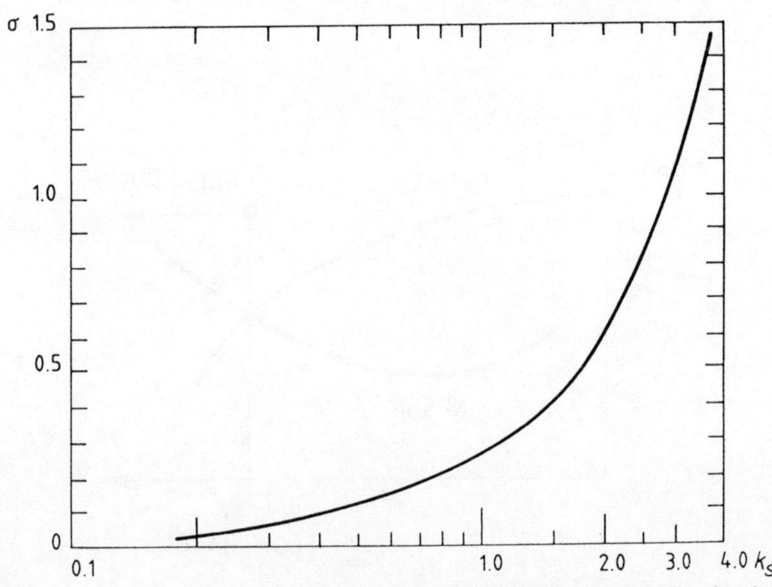

Fig. 5.7 A plot of the Thoma cavitation number against characteristic Number k_s

The Thoma number is defined as

$$\sigma = \frac{\text{NPSE}_R \text{ (Critical)}}{\text{Pump design specific energy rise}}$$

If $NPSE_A$ is larger than $NPSE_R$ given by this value, cavitation should not be a problem; if it is less the pump may be in trouble.

An alternative presentation is to use suction specific speed (or suction characteristic number), k_{SS}.

$$k_{SS} = \frac{\omega \sqrt{Q}}{(\text{NPSE}_R)^{3/4}}$$

and

$$k_{SS} = \frac{ks}{\sigma^{3/4}}$$

Typical values for k_{SS} are 3 for a middle range k_S pump, with good suction designs in the range 3.8–4.2; good condensate pump designs can go to 5.

5.3.2 For positive displacement pumps

For a reciprocating pump:

$NPSE_R$ is that value of NPSE at which specific energy falls by 3 per cent

Since positive displacement pumps are often used for liquids other than water, and flow is intermittent, general laws are difficult to state, and the references must be consulted for further data. Manufacturers should be consulted where high lifts are considered, and the literature consulted for more information.

5.4 NOISE CRITERIA FOR ROTODYNAMIC PUMPS

The UK National Engineering Laboratory of East Kilbride has published work done on large pumps that indicates (Fig. 5.8) a definite noise spectrum as the NPSE falls at constant flow and speed. The emitted noise reaches a maximum as NPSE falls, then reduces until the fall-off of gH at low NPSE occurs, when noise again increases. Parallel work on metal removal in a water tunnel indicates a close relation of maximum metal loss with high noise level. It is thus argued that surface damage occurs when noise level is high (Fig 5.8).

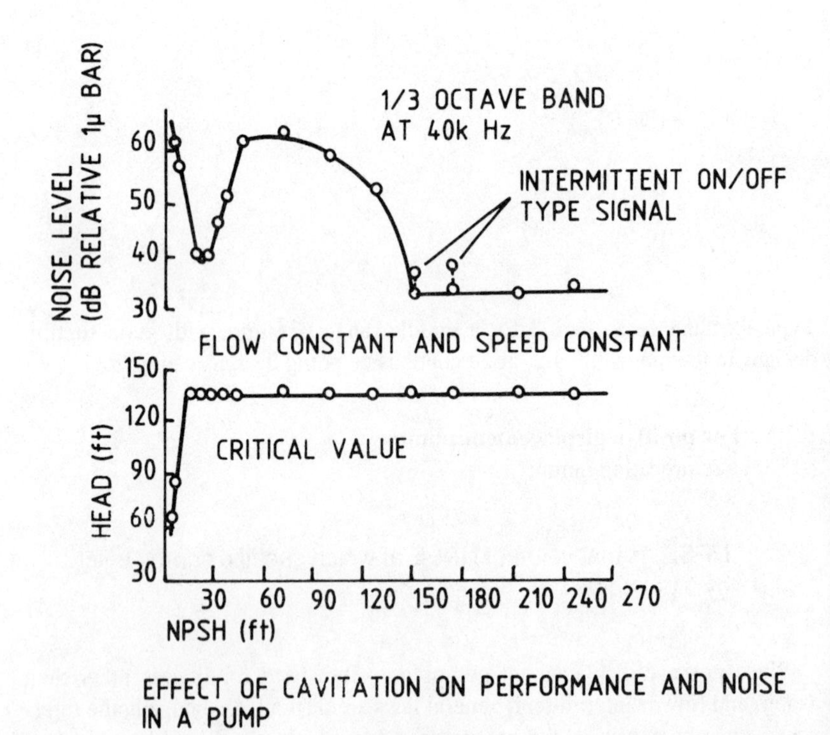

Fig. 5.8 Noise generated by a pump plotted against NPSH, based on work by Deeprose *et al.* (13)

5.5 SIZING OF SUCTION TO GIVE GOOD $NPSE_R$ FOR CENTRIFUGAL PUMPS

The main area subject to cavitation effects has already been identified as the suction zone. A number of pump studies to determine the optimum area to reduce cavitation problems have been performed, and the reader is referred to the References and the Bibliography. Pearsall **(14)** analysed available data and proposed a correlation for the optimum suction diameter. There are a number of problems, however, as the suction size is related to the onset of recirculation, as too large a diameter brings the critical point of breakdown closer to design flow, as illustrated in Fig. 5.9.

5.6 THE EFFECT OF HOT LIQUIDS

There is an improvement in the critical $NPSE_R$ as the liquid increases in temperature, so a pump tested on cold water which has a satisfactory cavitation performance will have an even better one if hot water is pumped. The reader should consult the References and the Bibliography for more information.

5.7 AVOIDANCE OF CAVITATION PROBLEMS

5.7.1 In the centrifugal pump

If a pump $NPSE_R$ is too high for a duty, and increasing the suction diameter is impractical, an inducer may be fitted. This device is fitted in the suction as shown in Fig. 5.10 and becomes effectively, an axial first stage. Typically the $NPSE_R$ curve can be improved as shown in Fig. 5.11. The hydraulic penalty is increased power requirement and, hence, lower efficiency. There is a complementary problem, in that the inducer and impeller can only be matched at one flow rate, imposing extra losses away from that operating point.

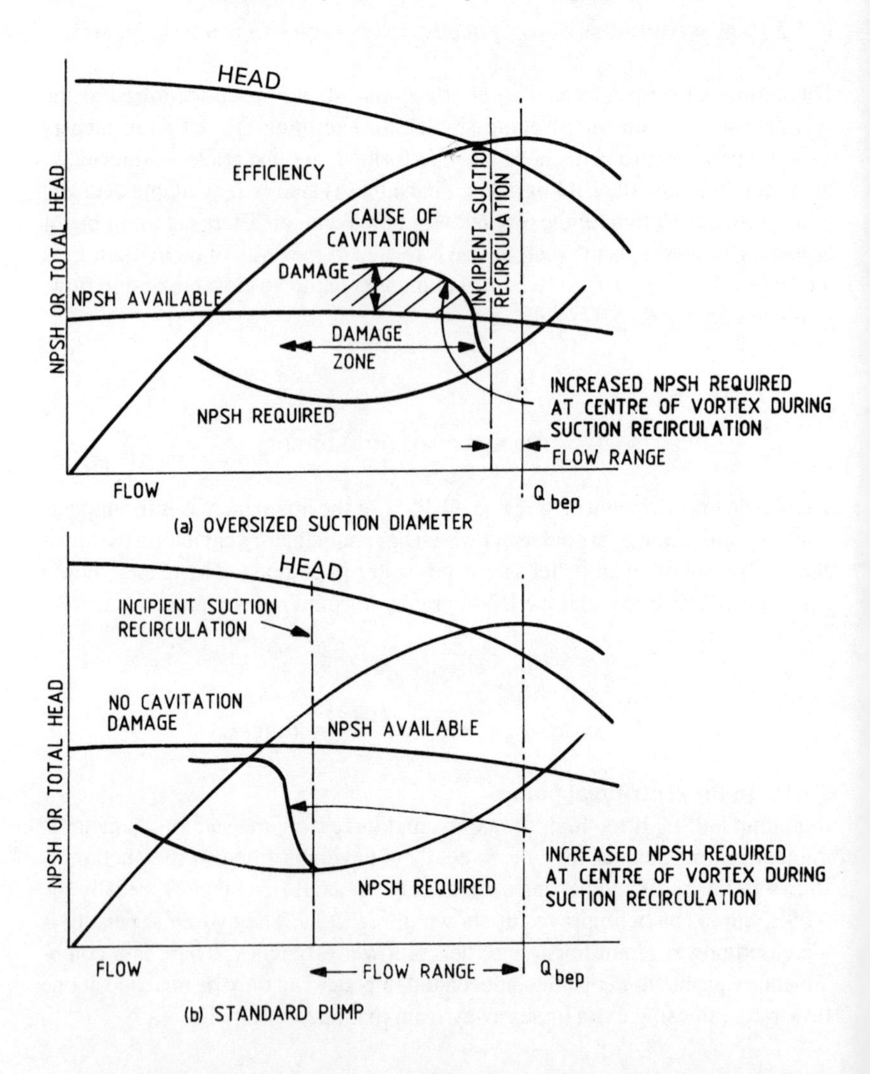

Fig. 5.9 The effect of suction diameter on the onset of recirculation and the effect on $NPSH_R$

5.7.2 Positive displacement pumps

With gear pumps the suction capability can be improved by improving entry to (and exit from) the pockets formed by the tooth profiles. The teeth are, therefore, not formed as those used in load bearing gears, and great care goes into the profiles, which in some designs differ on the front and rear flanks. In other pumps careful attention to clearance control may improve matters.

In reciprocating pumps increased suction pipe sizes and pulsation dampers are practical methods of avoiding cavitation problems.

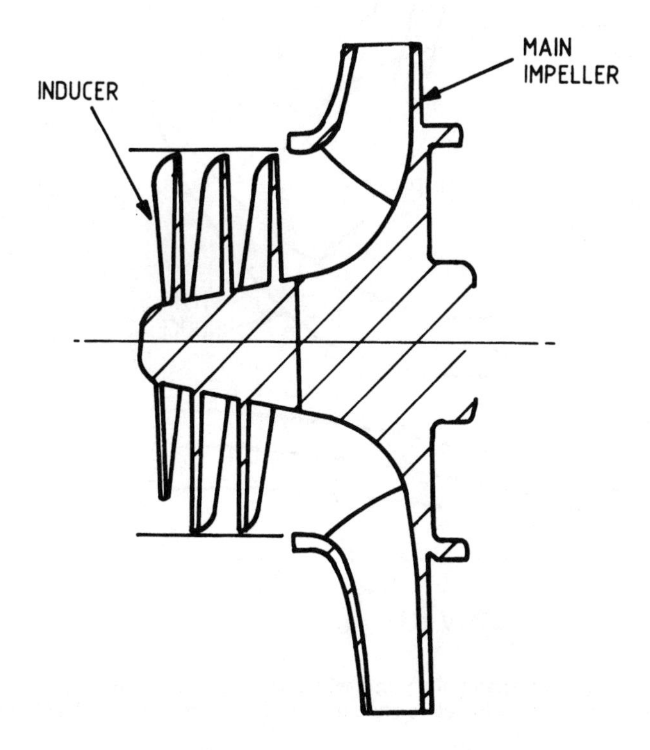

Fig. 5.10 The combination of an inducer and a centrifugal impeller

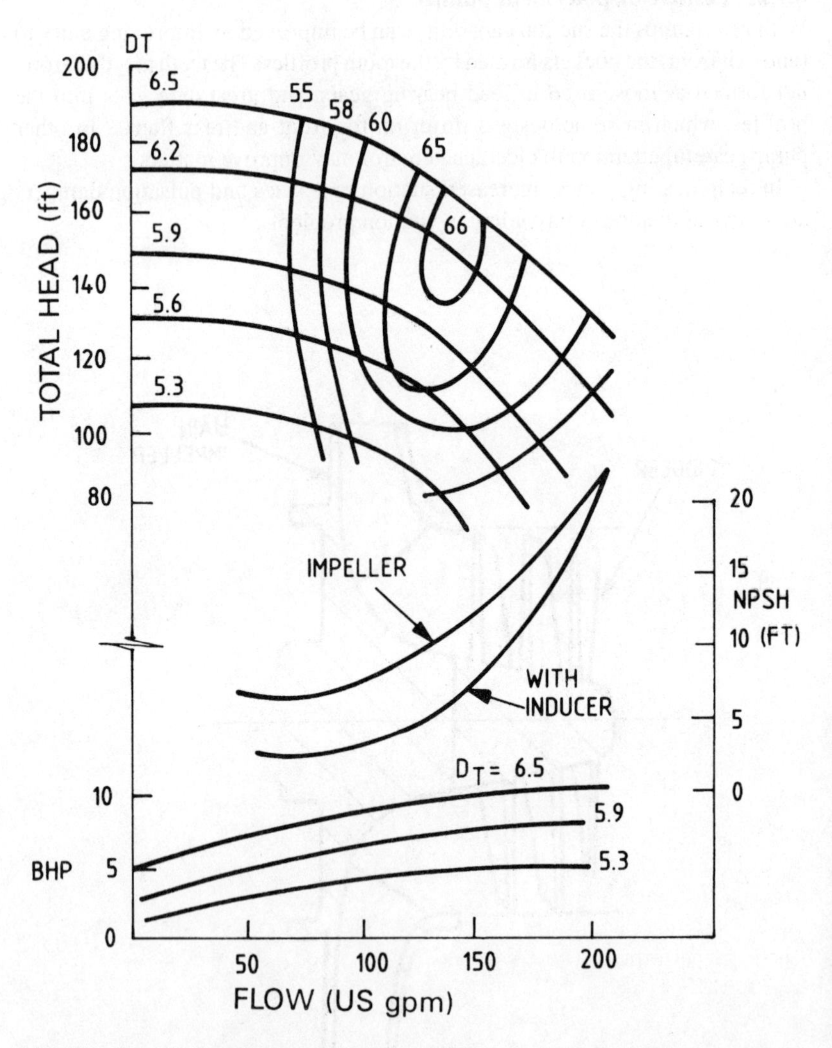

Fig. 5.11 The characteristic of a pump with and without an inducer. (a centrifugal pump fitted with a 2.5 in suction and a 1.5 in discharge, driven at 3550r/min)

CHAPTER 6

The Interaction between Pumps and Systems

6.1 INTRODUCTION

The methods used to calculate the losses due to flow in systems were discussed in Chapter 4. Since most systems need a pump to overcome flow resistance it is necessary to discuss how pumps and systems interact both in the steady state and in a transient way. The basic principles are discussed and the methods of adjusting pump output if system resistance changes are required, are outlined. Since some changes take place rapidly, transient problems will be described. The chapter closes with a discussion of starting pumps, which outlines good practice.

6.2 THE STEADY STATE INTERACTION BETWEEN A PUMP AND A SYSTEM

The flow losses in a system can be represented as shown in Fig. 6.1 with a static component and a dynamic loss. A long pipeline with many bends will have a large dynamic component, and a drainage scheme with a low lift will have a low static lift characteristic as shown. The constant speed characteristics of a centrifugal pump and a positive displacement pump are shown in Fig. 6.2, and, as can be seen, they do not conform to the typical system loss curves already discussed. If the two sets of characteristics are superimposed, as in Fig. 6.3, there is only one flow rate at which they cross; this is called the match point, and the flow rate at the match point is hopefully the duty flow for the system. If the flow is not that required for the system, adjustments either to the system or the pump involved need to be performed. These will now be described.

6.3 FLOW ADJUSTMENT BY SYSTEM CHANGE

If the pump is running at constant speed there are several system changes open to the engineer. The simplest is the use of valve opening or closure (Fig. 6.4). With

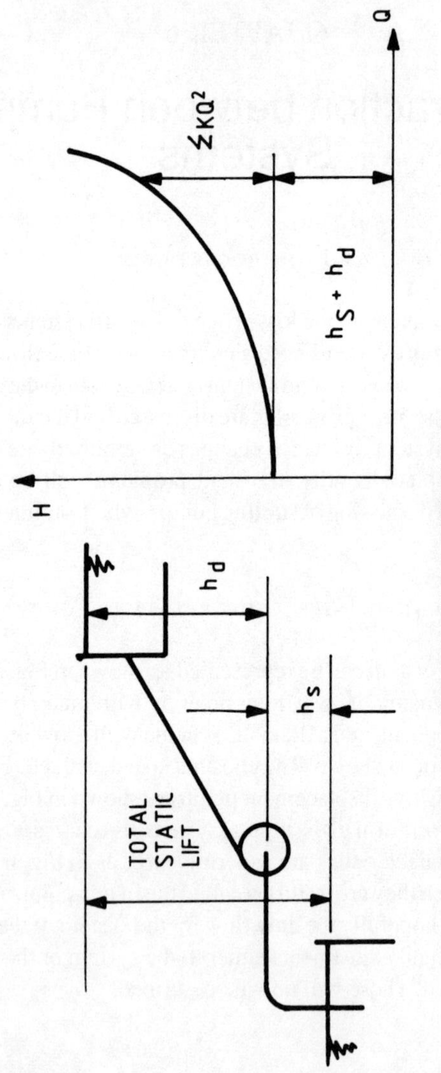

Fig. 6.1(a) Typical systems and their characteristics. Static lift and flow resistance

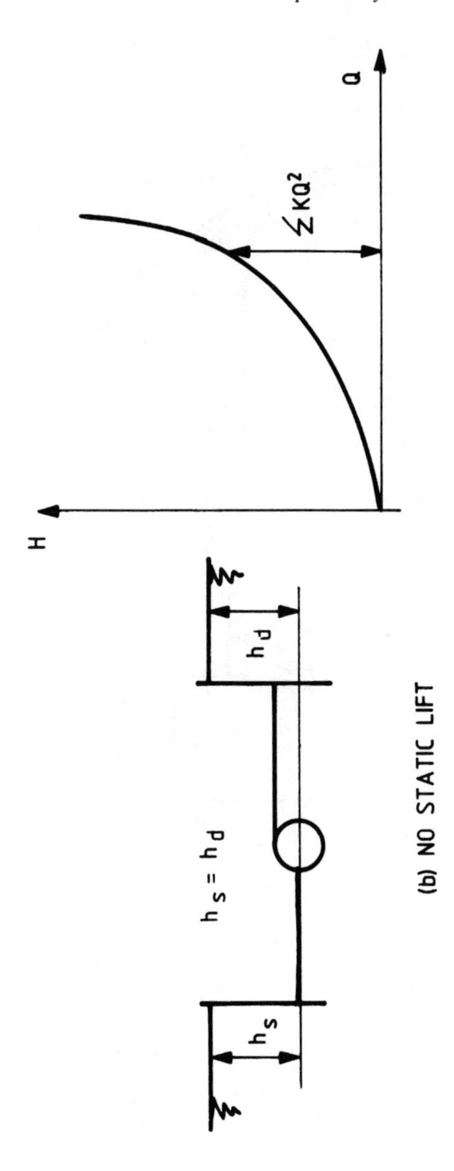

Fig. 6.1(b) Typical systems and their characteristics. No static lift

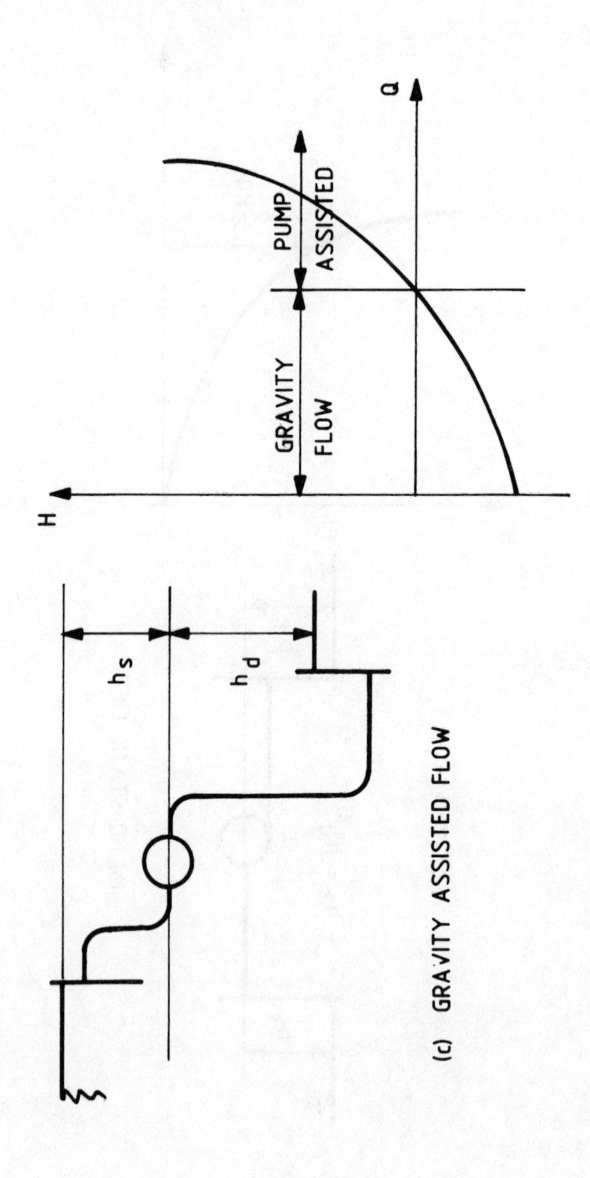

Fig. 6.1(c) Typical systems and their characteristics. Gravity assisted flow

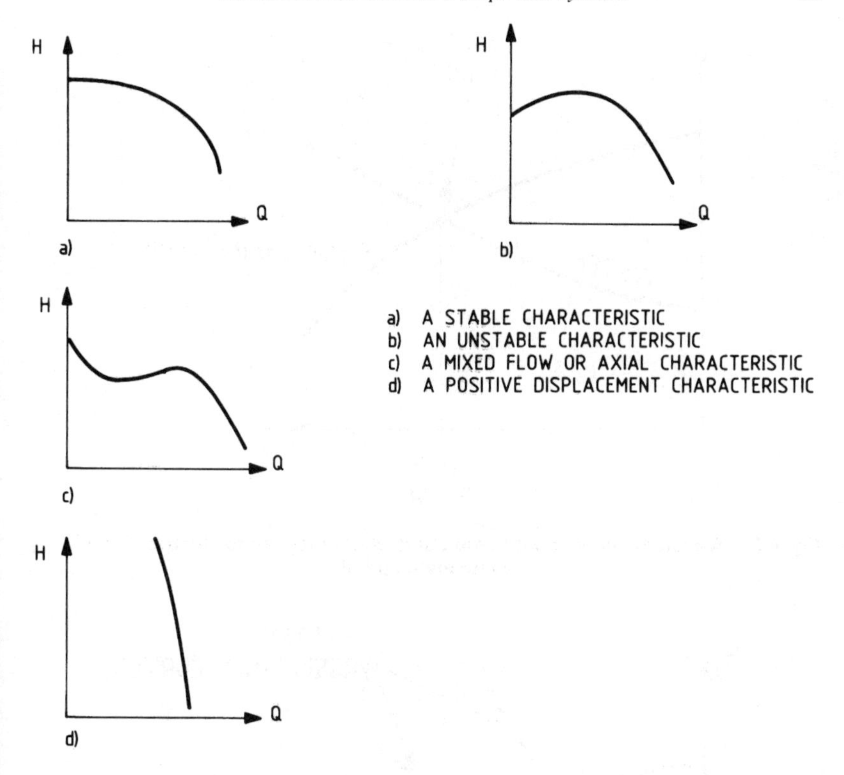

Fig. 6.2 The characteristics of a number of rotodynamic pump designs, and of a positive displacement design, compared

a centrifugal pump this gives a good flow range, as shown, but with a positive displacement pump the change of flow is very small. In both cases the control is stable, but at the expense of pumping power.

A system used with hot liquids, such as condensate, is to provide a by-pass to the suction well, so that while the pump runs there is a controlled flow back to the sump (Fig. 6.5). When the pump flow is stopped there is still a flow back and the risk of vapour locking due to the added heat in the pump case and impeller causing flashing to steam is avoided. This clearly reduces the volumetric

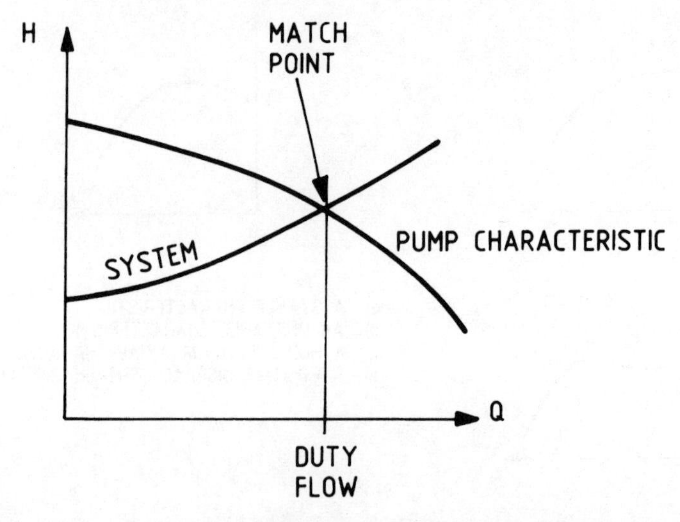

Fig. 6.3. A system curve and a pump curve superimposed to illustrate the steady state match point

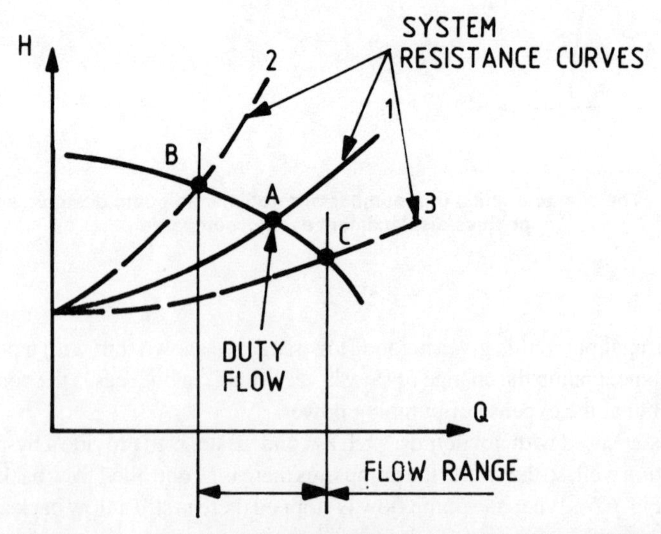

Fig. 6.4 System flow rate control using a discharge valve with a constant speed centrifugal pump

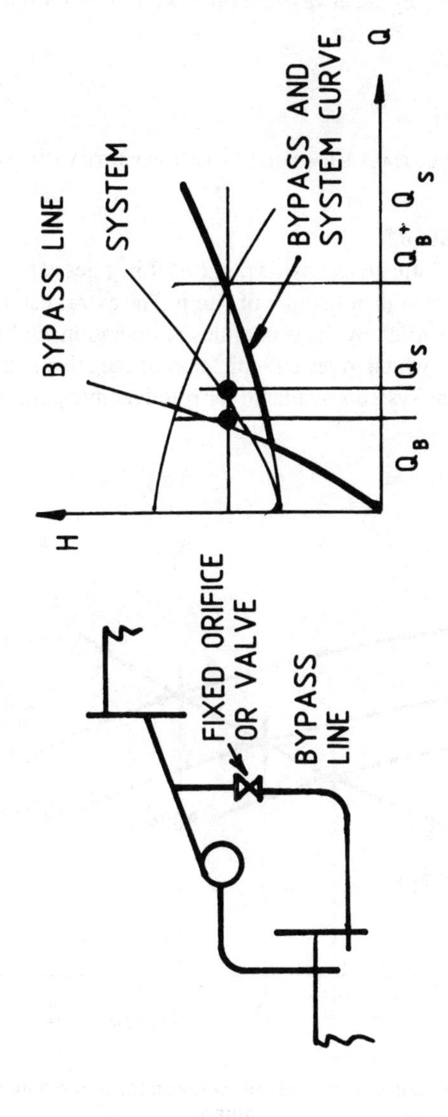

Fig. 6.5 System flow control using a by-pass system using a constant speed centrifugal pump

efficiency of the machine as a penalty. Positive displacement pumps are not at the same risk since they are always provided with a relief-valve which can be set at a suitable level.

6.4 FLOW ADJUSTMENT BY THE USE OF GEOMETRY CHANGES OR SPEED CONTROL

6.4.1 Speed control

As Figs 6.6 and 6.7 illustrate, a wide range of flow rates can be achieved by the use of speed control with both types of pump. The extra cost of the drive and its control must be justified by the economics of operation, and Bower (**15**) did a comparative study which gives an indication of the advantages and disadvantages of the various systems available for rotodynamic pumps.

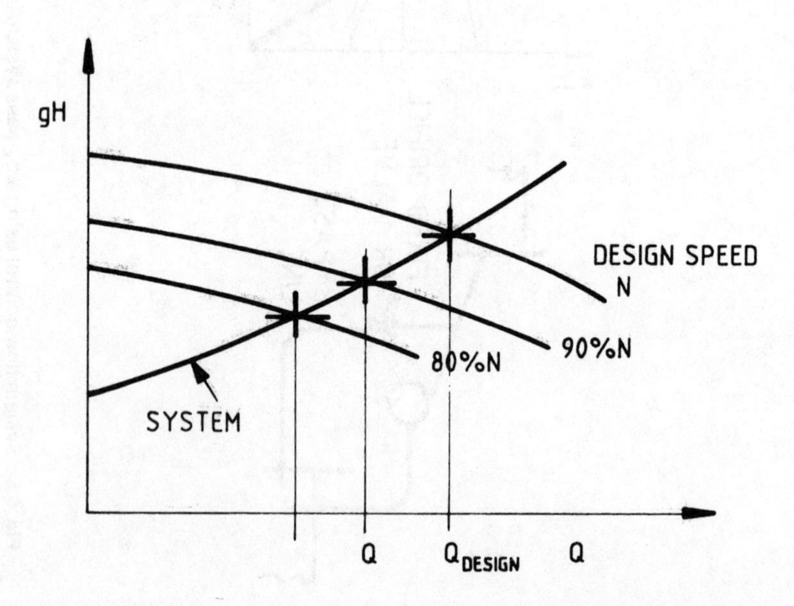

Fig. 6.6 Flow rate control using speed variation for a constant speed centrifugal pump

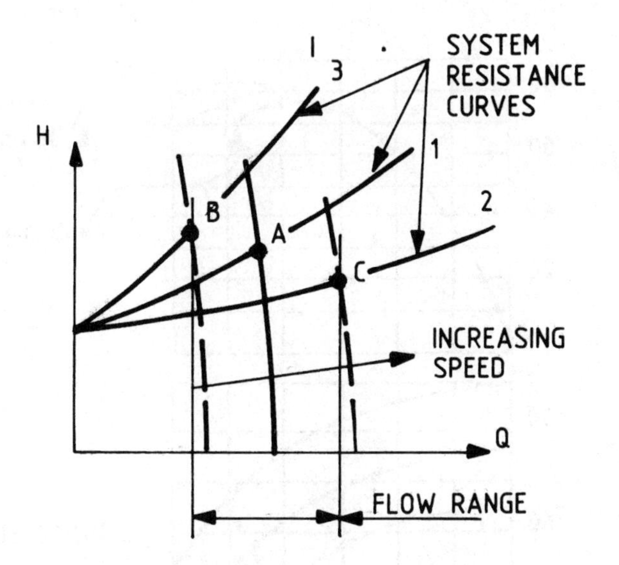

Fig. 6.7 System flow control using a variable speed positive displacement pump

6.4.2 Geometric changes

There are several geometric changes available to the centrifugal pump engineer; diameter change, impeller tip width alteration, changes in outlet angle, and, for small changes in flow or head adjustments, changes in the shape of the outlet tip of the impeller blades. All of these are permanent and must be used with care.

Diameter change is much used as it allows a standard impeller to be produced and the machine output changed to suit a customer specification that is within 15 per cent of the standard duty. The adjustments are well covered in the *Pump Handbook* (**16**): they follow the Scaling Laws, usually written

$$D_1/D_2 = (gH_2/gH_1)^{1/2}$$
$$Q_1/Q_2 = (gH_2/gH_1)^{1/2}$$
$$P_1/P_2 = (gH_2/gH_1)^{3/2}$$

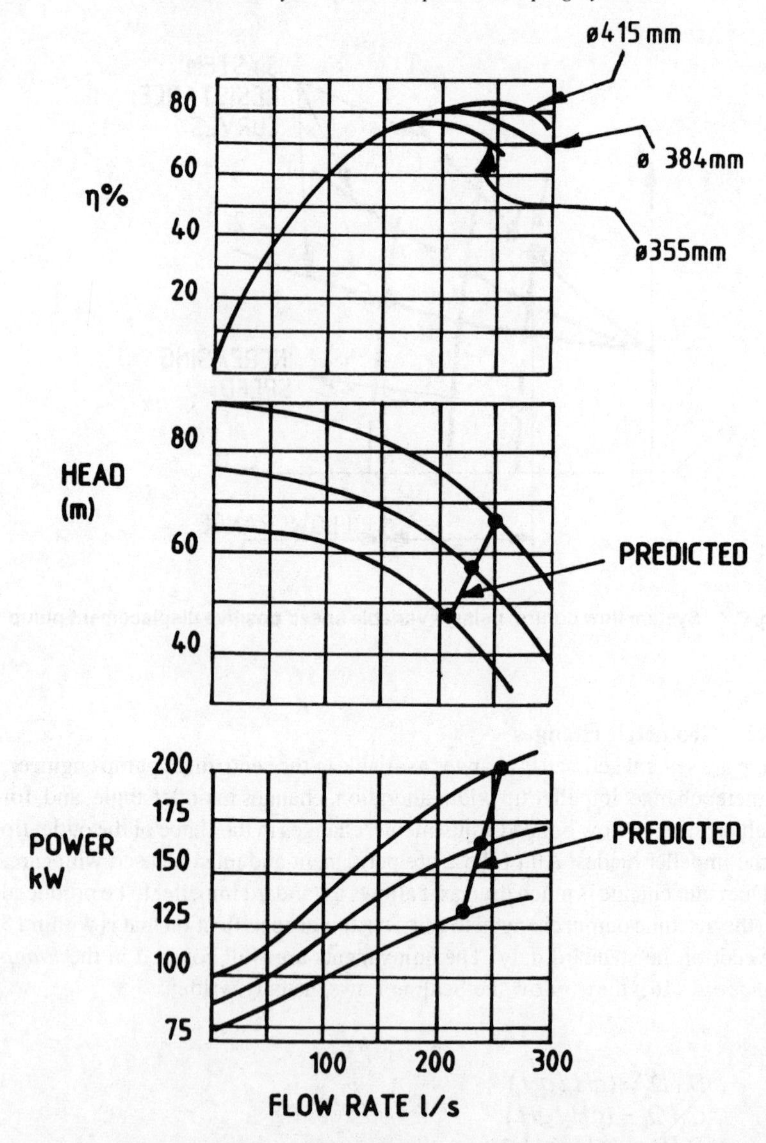

Fig. 6.8 An example of the use of the scaling laws to predict the effect of reducing impeller diameter for a centrifugal pump

Thus, to alter the head, the diameter is changed, and the change in flow rate and (approximately) the power can be estimated. Conventionally, a diameter reduction of up to about 15 per cent can be made without departing from the principle of 'dynamic similarity' on which the scaling laws are based. Power tends to change more than the prediction because flow is not strictly similar and losses vary. Figure 6.8 illustrates this with an actual example.

Larger changes available are alterations to the tip width and modifications to the outlet angle (Fig. 6.9). The latter allows changes to the flow range and to the slope of the head-to-flow curve, as sketched. Both however are major redesigns, and require consideration of the design of the casing which accepts the flow; this is discussed in other textbooks. An adjustment that is easy to make if the pump is not quite at the duty point required is to modify the outlet edge shape of each impeller blade. This is illustrated in Fig. 6.10, where the filing of the tips from the normal shape produced by turning the impeller improved the head rise by almost 8 per cent.

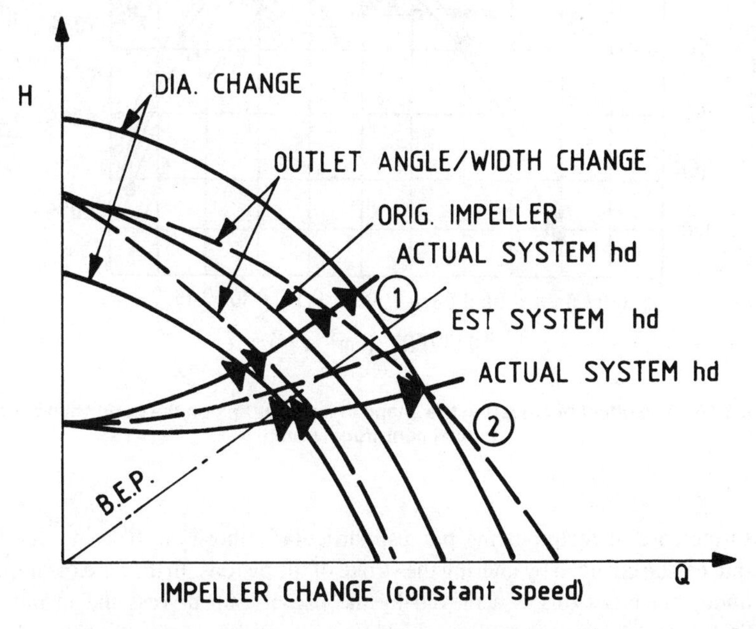

Fig. 6.9 The effects of changing diameter and tip width on the output of a centrifugal pump

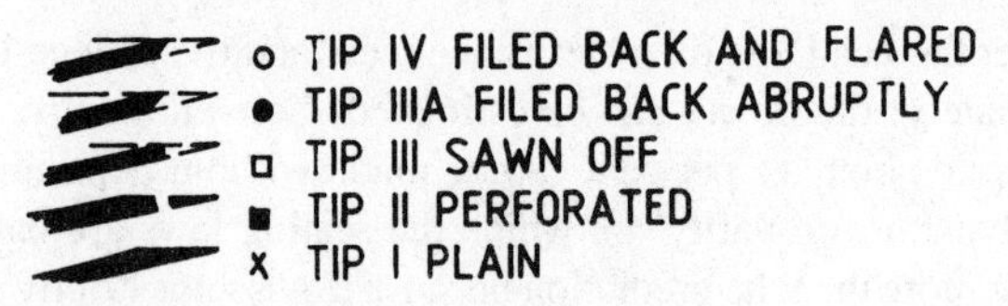

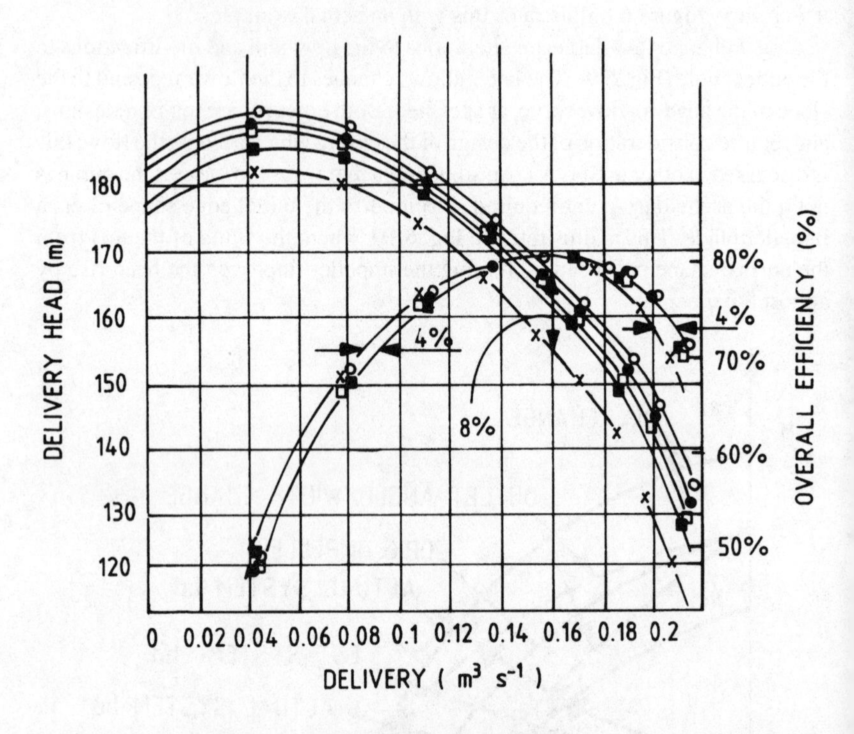

Fig. 6.10 The effect of adjusting the shape of impeller tips on the performance of a typical centrifugal pump

In the case of reciprocating pumps, particularly those on metering duties, output can be changed by altering the stroke of the pistons. In the case of single-cylinder machines this is achieved by the use of cam drives, and in multi-cylinder machines a swash plate which can be varied in angle may be used, as shown in Fig. 6.11.

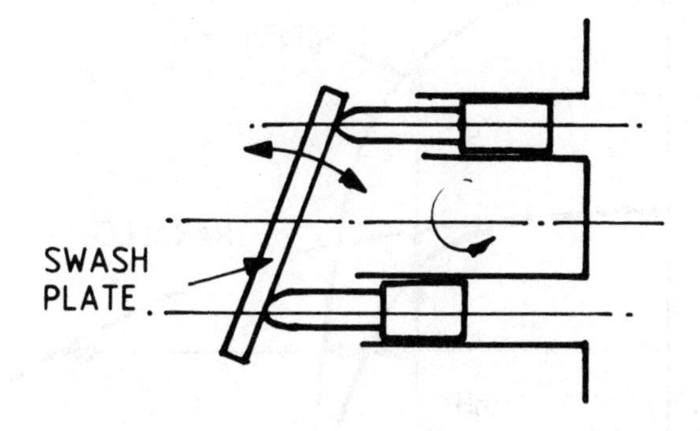

Fig. 6.11 A method of flow control in a reciprocating pump using a swash plate

6.5 PUMPS IN SERIES OR PARALLEL

Due to the fact that applications needing positive displacement machines rarely need pumps to be placed in series or parallel it is only necessary to discuss the ways in which rotodynamic pumps can be arranged in such a way. Normally pumps are identical, and the most common systems used utilise pumps in parallel to increase the flow through a system, and Fig. 6.12 demonstrates how the combined characteristic for two pumps is found. Also shown is the method of establishing the combined output of two pumps in series.

It will be noted that when pumps are in series the inlet pressure to the second pump is the delivery level from the first, so that it is subject to higher loads, and therefore, the seal rating must change also. When pumps are in parallel, non-return valves must be fitted in the delivery line so that when one is running it does not drive the other as a turbine.

As Fig. 6.13 shows, it is possible to place many pumps in parallel to allow for a range of flows to be delivered to the system as demand requires. The fact that the system curve follows a square law means that if three pumps are running the flow is not three times that obtained with one pump operating. Figure 6.14 illustrates the further flexibility conferred by providing the pumps with variable speed drives. In practice, pump stations will be provided with one variable speed machine to provide flow control between the levels provided by pumps running

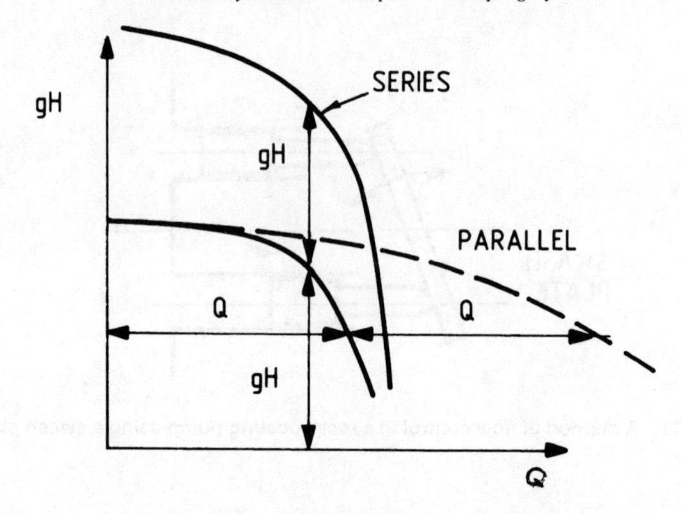

Fig. 6.12 The combined characteristics formed by centrifugal pumps in series and parallel

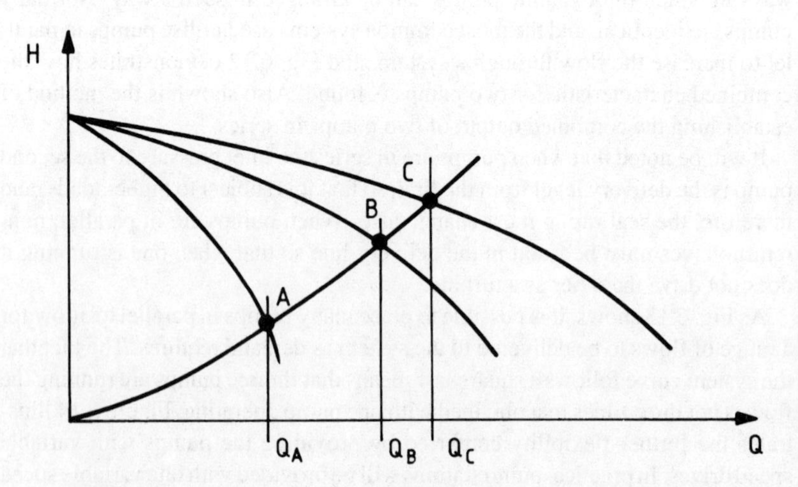

Fig. 6.13 The effect of operating three identical pumps in parallel

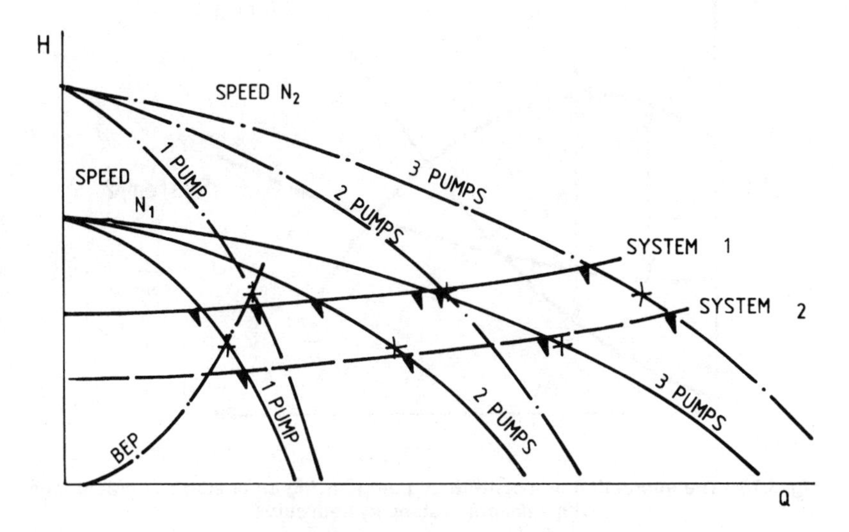

Fig. 6.14 The way that three pumps operating in parallel and at different speeds interact with two system curves

at constant speed as they are turned on; this gives an economy in operation and first cost for the station. Very rarely are dissimilar pumps combined because they tend to give some instability in operation.

6.6 A DISCUSSION OF STARTING AND CHANGING OF DUTY POINT

6.6.1 Stability of systems

It has been assumed in the situations described that the operation is stable. Using Fig. 6.15 which shows two systems supplied by a pump which has a characteristic which reaches a maximum head and then falls to 'shut valve'. (Such a machine is called an unstable pump). If the pump is supplying system (1) the excess head over the system demand provided by the pump is h_1, which reduces from 'shut valve' to the match point, and thus flow will increase steadily and with stability. If the pump is supplying system 2 the difference is h_2 which increases from 'shut valve' to point A, from which it tends to decrease towards

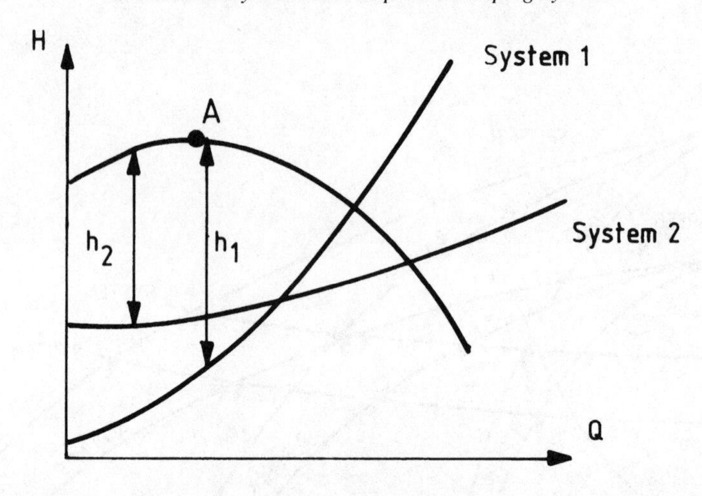

Fig. 6.15 The interaction of a centrifugal pump having an unstable characteristic with a flat and a steep system curve

the match point. This will need manipulation of the system valving to make up the difference, and the probability is of instability. It is desirable that the use of pumps with unstable characteristics be avoided.

6.6.2 The implications of electric motor characteristics during starting

Two forms of starting are in common use; direct on-line, and Star-Delta starting. When a radial flow pump is started the pump torque will always be less than motor torque when started direct on-line. The worst case is that of starting on open rather than shut valves. With an axial flow pump starting on a closed valve the pump demand is larger than motor torque as full speed is approached, as Fig. 6.16 illustrates. Also shown in the figure is the pump demand for a high static lift system on open valve. Figure 6.17 shows how pump demand and motor output match during Star-Delta starting. Care is needed to check what the motor curve is like in Star-Delta to avoid the risk of stall during the start cycle.

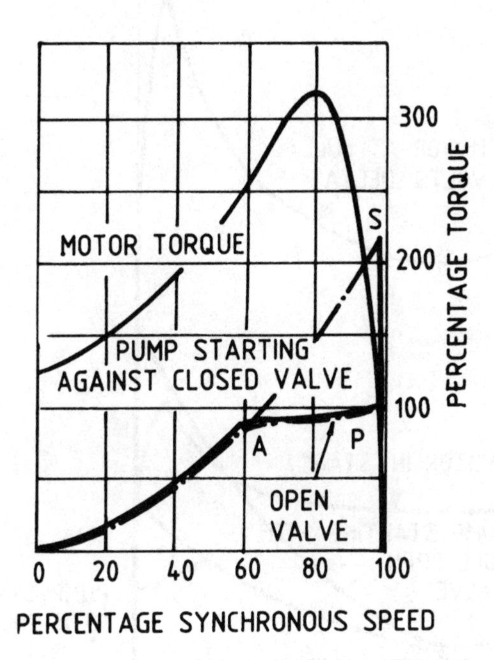

Fig. 6.16 Starting an axial pump with an electric motor

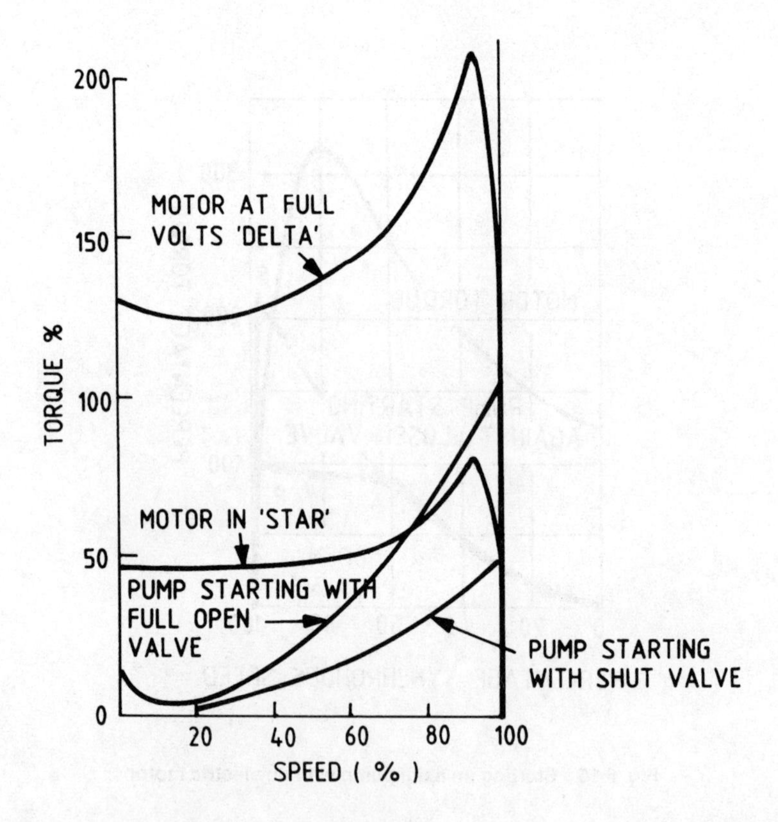

Fig. 6.17 Starting a centrifugal pump driven by an electric motor using a Star-Delta starter

CHAPTER 7

Selection and Application of Pumps

7.1 INTRODUCTION

When a project engineer is selecting a pump for a new or refurbished plant, a number of basic criteria must be satisfied. The pump must:

- be the correct type;
- deliver the duty or duties specified;
- be constructed of the correct materials to withstand the fluid pumped;
- be offered at the right price;
- be economical to run and maintain.

Additional criteria which are related to the process will also need to be satisfied.

In earlier chapters the basic principles of pumps and the way they interact with systems have been discussed. In this chapter the general principles which guide the correct selection and application of pumps will be discussed. The effects of the dimensional standards and industrial codes will be described.

7.2 ESTABLISHING THE PUMP DUTY

It is very important to provide a statement of the range of flows and pressure rises as well as a duty point for focusing the best efficiency flow for the pump. Processes frequently run over a range of flows and in some cases the temperature and, hence, fluid viscosity will vary. It is, therefore, important that the pump maker is aware of the range of duties so that the most onerous operating duty can be allowed for, and the correct material and configuration selected.

Industry codes such as API 610, which will be discussed below, were partly developed to encourage and formalise the interchange of information between

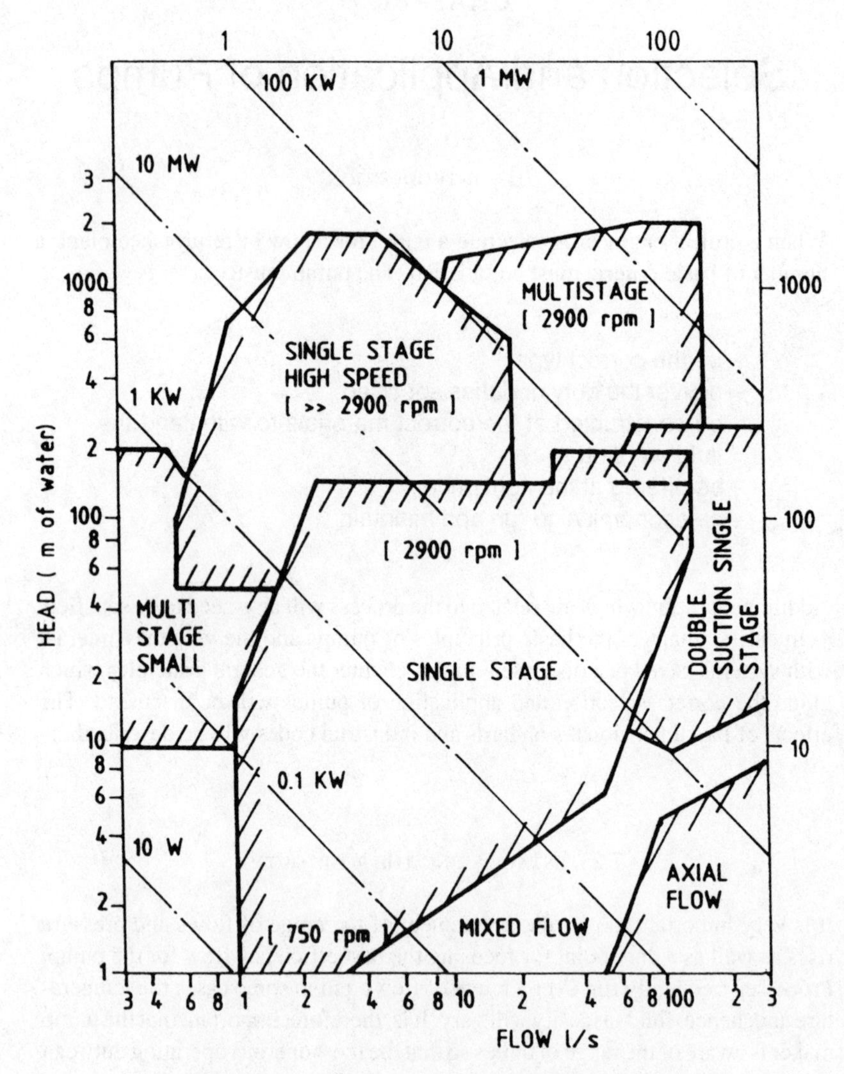

Fig. 7.1(a) Selection chart for rotodynamic pumps

the customer and the supplier of the pump, and comprehensive data sheets which contain a wide range of information have been developed.

7.3 WHICH TYPE OF PUMP?

Having established the duty and fluid data the pump type must be established. In most cases the flow rate and energy rise will determine the type of pump. For example, if Figs 7.1(a) and (b) are consulted, a pump to deliver 10 litres per second with a head rise of 100m will be a single stage centrifugal machine running at 2900 r/min, with a power input of about 10kW.

In a number of cases a centrifugal and a screw pump will give the same duty, with the latter having a superior $NPSE_R$. Decisions then must be based on company experience. Large companies like the petrochemical and chemical multinationals have their own established procurement procedures. A typical summary of such a procedure for duties where there is a possibility to choose would be that an end suction centrifugal pump shall be selected. (This will probably be over 80 percent of the duties.) If a centrifugal machine is not suitable, a positive displacement pump will be used. Some companies prefer rotary positive displacement pumps, except where metering accuracy means a reciprocator has to be selected.

7.4 EFFECT OF FLUID PUMPED ON PUMP DESIGN

7.4.1 Viscous effects

If the fluid has a high viscosity the reduction in performance this causes for a centrifugal pump must be remembered. A screw pump will offer a better performance with higher input power, though volumetric efficiency improves. Screw pumps can pump fluids with viscosities up to 1500 cSt. At higher viscosities gear pumps should be used; they have pumped liquids with kinematic viscosity of 25 000 cSt.

7.4.2 Dirty and aggressive liquids

Many process duties involve highly aggressive chemicals, so that great care is needed in material selection. Table 7.1 lists some materials commonly used for aggressive duties. A common design provision is to provide a double thickness

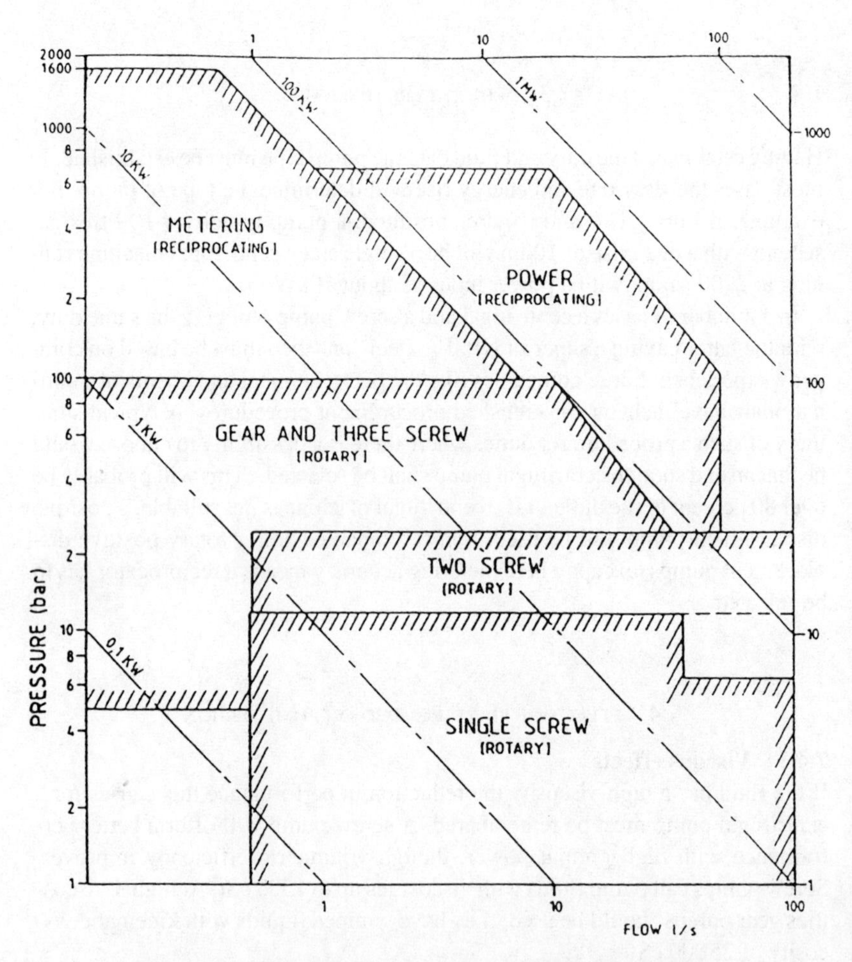

Fig. 7.1(b) Selection chart for positive displacement pumps

of casting to allow for material loss but still ensure pressure integrity for a service period.

Table 7.1 Some commonly used materials for chemical pumps

Liquid	*Material*
Water, caustics at low temperatures, solvents	Cast iron to BS1452 grade 17 or 18/8 stainless steel
Nitric acid, caustics, xylene, toluene, and other solvents	18/8 stainless steel
Caustics, solvents, general chemicals, acetic acid, hydrogen peroxide	18/10/Mo stainless steel
HCL, barium chloride, sodium hypochlorate	Hard rubber lining
Sodium hypochlorite, chlorinated brines	Titanium
Aluminium chlorohydrate, HL, thorium sulphate	Epoxy resins

Alternative designs offered for aggressive liquids are lined with inert materials such as teflon, or offer simple shapes using ceramics, as shown in Fig. 7.2.

Solids-handling designs such as machines which pump solids in suspension (gravel or limestone for example) or sewage pumps, have to be able to pass large-size solids. A test requirement for a sewage pump, for example, is that a 100mm sphere shall pass through the impeller. A typical design is shown in Fig. 7.3, and Fig. 7.4 shows impeller designs in common use.

Reciprocating pumps are also used in solids-pumping systems, with valves specially developed to be self-cleaning. Peristaltic designs have been developed which will pump gravel using armoured hose enclosed in an oil bath to provide lubrication and cooling.

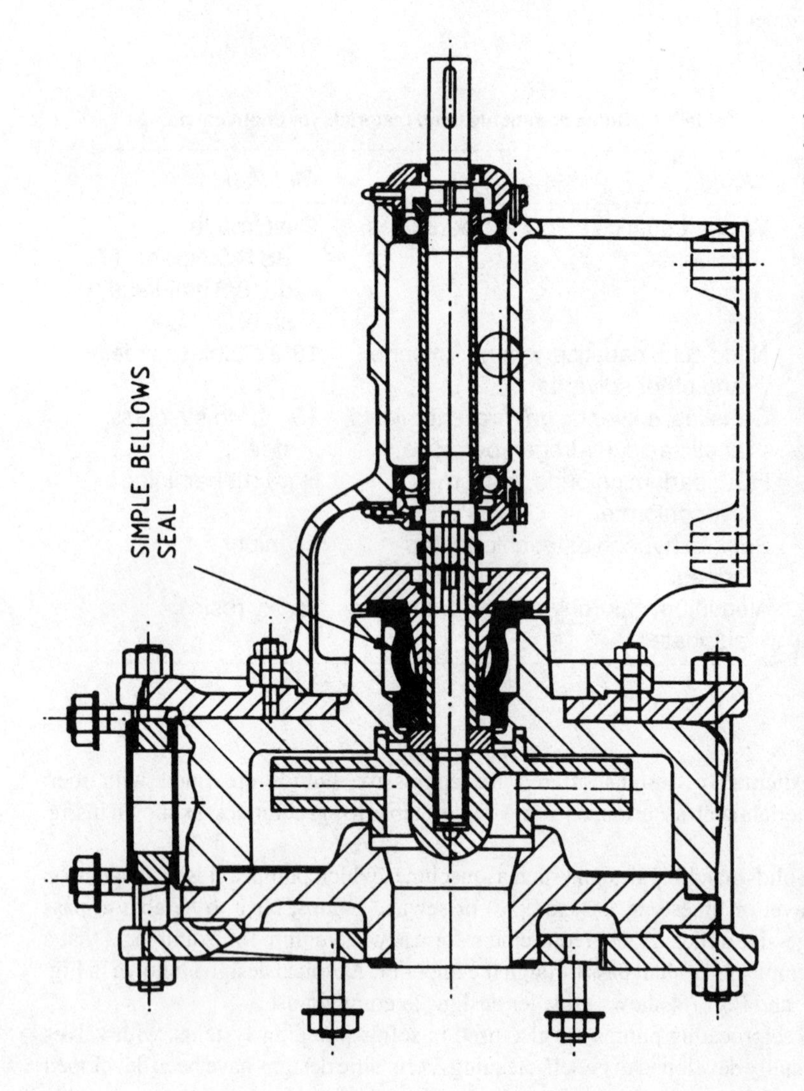

Fig. 7.2 A chemical pump constructed from ceramic materials provided with a mechanical seal

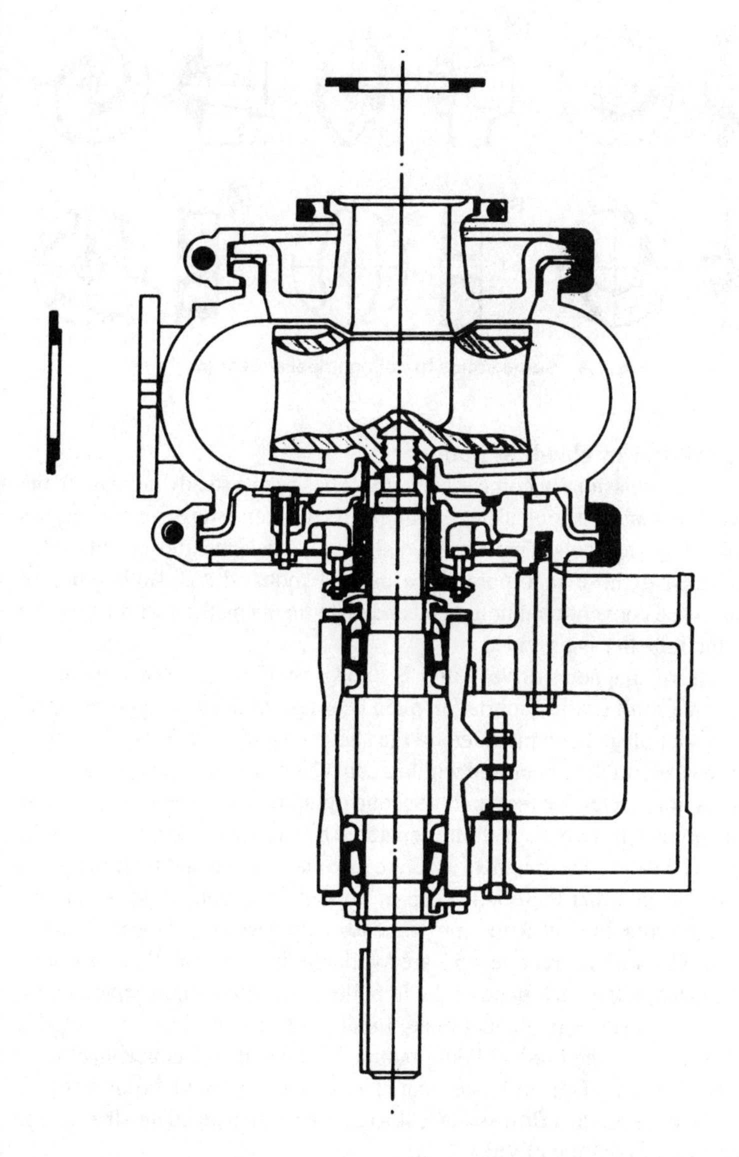

Fig. 7.3 A typical gravel pump

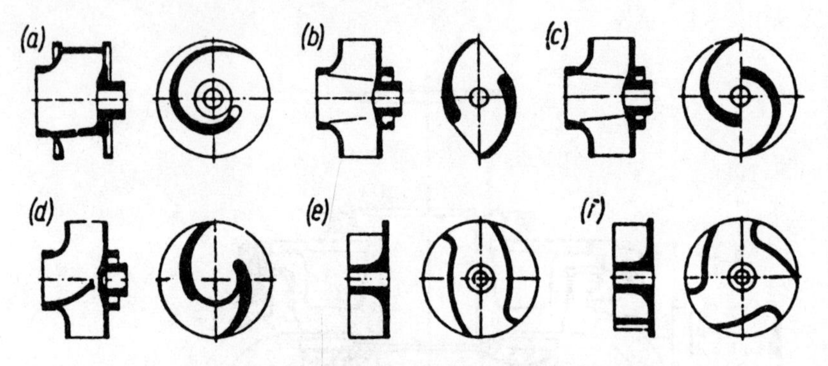

Fig. 7.4 Some solids handling impeller designs

7.4.3 Leak-free or glandless pumps

As controls on emissions become stricter, there has been a steady increase in the application of glandless (or leak-free) pumps. As the name suggests, these types of machine have no seals. There are two alternative designs, those using magnetic drive, and those incorporating a canned motor drive. Both types of machine have a conventional liquid end and differ in the method with which the drive to the impeller is provided.

One type of magnetic drive pump is shown in Fig. 7.5. The impeller is mounted on a short shaft supported in plain bearings which are supported from the pump back plate. Also mounted on the shaft is a cylindrical magnetic ring. This rotating assembly is enclosed within a can which provides the enclosure for the process liquid. Outside the can is a second cylindrical magnet ring driven by the drive motor. The two magnets are separated by two small clearances and the thickness of the can. The magnets used are rare earth based and their magnetic interaction is such that there is no slip in normal operation. In some cases a second liquid containment shell is provided to contain leakage if the main can is punctured. Several different designs are available; in some small pumps magnets are placed in the back plate of the impeller and a diaphragm separates the impeller from the driving magnet plate. In all cases the bearings are provided with lubrication by the product being pumped, which also conducts heat from the inside of the can. This means that care has to be taken about the siting of the fluid return to the product flow system, as a hot liquid introduced into the suction could provoke cavitation onset.

A canned motor is shown in Fig. 7.6; as can be seen the liquid end is mounted on the motor shaft. The stator is provided with a sleeve which provides an inner can to keep the liquid circulating from the liquid end to the thrust bearing at the upper end of the assembly. The bearings are journal type and the product passes through them to provide the necessary cooling and lubrication.

Both types of machine need to be designed to ensure that liquid is always present in the drive, and need the provision of monitors to ensure that the can or the motor are not full of gas.

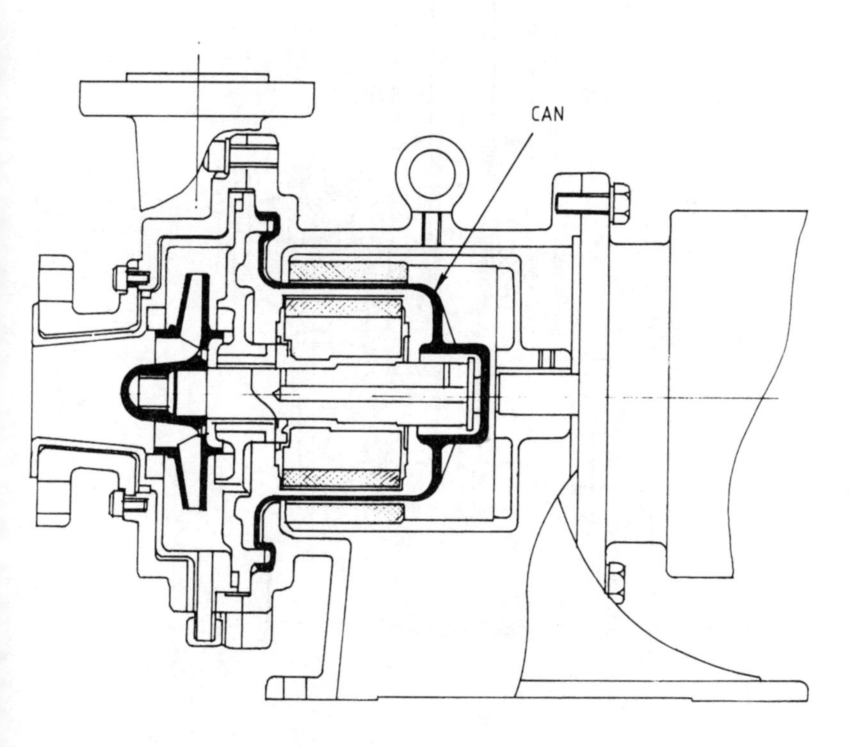

Fig. 7.5 A magnetic drive pump

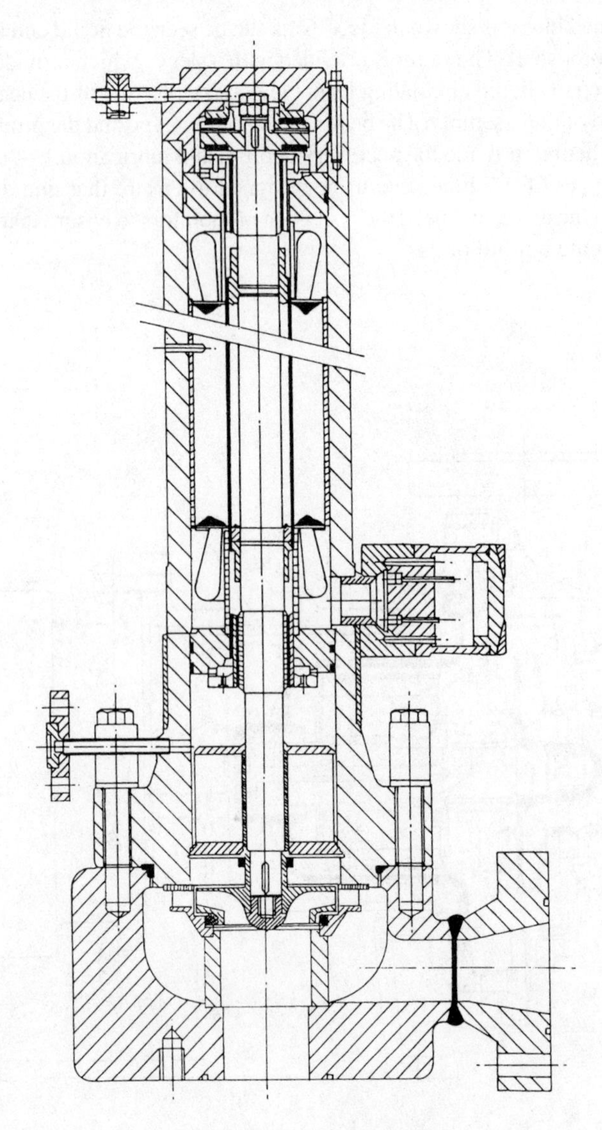

Fig. 7.6 A canned motor pump

7.5 PRIMING

The pumping action of a pump depends on the suction line and machine being full of liquid. Positive displacement pumps will pump a mixture of liquid and gas, though with a lower volumetric efficiency, but rotodynamic machines will only cope with gas contents up to about 10 per cent, and if this proportion increases the pump will lose 'prime' and 'vapour lock', with air filling the impeller passages which are then incapable of expelling the gas. Pumping action ceases and the pump then has to be primed, that is, refilled with liquid. If the pump is 'drowned', that is, the suction tank level is above the pump, the re-priming process takes place easily and flow can re-start.

In many installations the pump has to provide a suction lift, or, as in the tanker off-loading situation, draws through a line which is initially empty. In the first case the suction line will not empty if a foot-valve or non-return valve is fitted and the liquid is not volatile, and thus pump re-start is not a problem. In the latter situation a rotodynamic pump needs assistance in priming, and a positive displacement pump, with wetting of the rotors when needed, will evacuate the air and begin delivery. Ship cargo pumps are thus often scew machines or other high-flow positive displacement designs, as they require no other equipment for priming.

Two alternative methods are employed with rotodynamic machines (usually centrifugal): an external priming device, or integral means (known as self-priming designs). External devices available in common use are liquid ring pumps or jet ejectors, which draw air through the suction line and pump, and are disconnected when the pump is primed and begins pumping. As an example of the time to prime, fire pumps on fire appliances will prime using a liquid ring pump in two minutes or less, depending on the length of the suction line. Internal devices provided consist of a specially designed casing which retains a quantity of the pumped liquid when the pump is shut down. This liquid is then circulated by the impeller from the suction to the delivery chamber when the pump starts. This circulation of liquid entrains air from the suction line and discharges it down the delivery pipe until priming is complete. In this type of system the re-circulation continues during pumping with most designs, and is thus a source of loss as the volumetric efficiency is reduced.

7.6 THE INFLUENCE OF STANDARDS AND CODES OF PRACTICE

An important factor in selecting pumps is the influence of international, national, and industrial codes. The international and national codes define leading dimensions, methods of testing, and materials specifications, and they form the basis of contracts between customer and supplier.

Industrial codes like API 610 are all embracing. This code relates to centrifugal pumps, and has been developed by the American Petroleum Institute in consultation with companies involved in the petroleum industry. The code is always being updated and extended, at intervals of three to five years, in the light of developing technology. It is a comprehensive industrial code which is designed to provide a safe pump fit for duty in hazardous refinery and petrochemical applications. It covers all aspects: bearing and seal systems, casing provision, shaft design, vibration and noise, and running clearances; similar codes cover positive displacement pumps.

An important provision in such a code is a comprehensive questionnaire that forms the basis of the contract between customer and supplier. This covers the range of flows and head to be met, the ranges of fluid properties to be expected, the installation conditions, and other related data. The pump maker has to provide data about the pump to be supplied. The object is to ensure a full and precise exchange of information between supplier and user so that the fitness of the pump for duty can be assured. Such codes quote standards like those sponsored by the International Standards Organisation (ISO) and by the American National Standards Institute (ANSI). These give dimensional envelopes for the pumps and also the way in which hydraulic data are to be obtained and supplied.

The Bibliography gives a selection of the relevant codes and standards. The reader is advised that since they are subject to revision and/or replacement by the issuing body, a check on the relevant material should always be made.

7.7 CONCLUDING COMMENTS

In a book of this length it has only been possible to introduce principles, and in places to give a little detailed practical information.

A selection of the wide range of pumps available has been made, based on usage in industry. The basic principles which relate to performance have been discussed, and they have general application. If greater detail is needed, the reader may consult the literature cited.

Development in the pumping industry has been steady for over a hundred years, with recent changes in material science and technology allowing improved accuracy in casting profiles and materials of greater integrity. Recent technological developments have been made in mechanical seals and magnetic drives for seal-less pumps, and in developments which allow higher driving speeds and consequent reduction in size, thus reducing the cost and the space occupied by the pump.

REFERENCES

(1) **J.D. Summers-Smith** (Editor), *Mechanical seal practice for improved performance*, Second edition, 1992 (Mechanical Engineering Publications, London).

(2) *Seal and sealing handbook.* Third edition, 1990 (Elsevier, London, Amsterdam, New York).

(3) **J. Davidson** (Editor), *Process pump selection: a systems approach*, 1986 (Mechanical Engineering Publications, London).

(4) **G.F. Wisclicenus**, *Fluid mechanics of turbomachinery*, 1965 (Dover, New York), Volumes I and II.

(5) **F.A. Varley**, 'Effects of impeller design and roughness on the performance of centrifugal pumps', *Proc. Instn mech. Engrs*, 1969, **175**, 955–989.

(6) **I. Stepanoff**, *Centrifugal and axial pumps*, 1976 (Wiley, New York).

(7) **H.H. Anderson**, *Centrifugal pumps*, Third edition, 1980 (Elsevier, London, Amsterdam, New York).

(8) **R.C. Worster**, 'The flow in volutes and its effect on centrifugal pump performance', *Proc. Instn mech. Engrs.* 1963, **177**, 843–876.

(9) **J.E. Miller**, 'Liquid dynamics of reciprocating pumps: 1, forces acting on the liquid, and NPSH explained; 2, pulsation control devices and techniques', *Oil and Gas Journal*, 1983, April; May.

(10) **D.S. Miller**, *Internal flow systems* (2nd Edition), 1990 (Gulf Publishing Company, Houston, London, Paris, Zurich, Tokyo).

(11) **P. Lush**, 'Design for minimum cavitation'; 'Materials for minimum cavitation', *Chart. mech. Engr*, 1987, **34**, September, 22–24; October, 31–33.

(12) **R.T. Knapp *et al.***, *Cavitation*, 1970, Engineering Societies Monograph (McGraw-Hill, New York).

(13) **W.M. Deeprose, N.W. King, P.J. McNulty, and I.S. Pearsall,** 'Cavitation noise, flow noise, and erosion', *Cavitation*, 1974 (IMechE, London) pp. 373–381.

(14) **I.S. Pearsall**, *Cavitation*, 1972 (Mills and Boon, London).

(15) **J. Bower**, 'The economics of operating centrifugal pumps with variable speed drive' in *Developments in variable speed drives for fluid*

machinery, 1981, (Mechanical Engineering Publications, London), pp. 55–62.

(16) **I. Karassik *et al.***, *Pump handbook*, 1976, (McGraw-Hill, New York).

BIBLIOGRAPHY

American Hydraulic Institute Standards for Pumps, Fourteenth edition, 1983, (American Hydraulic Institute).

API 610: *Centrifugal pumps for general refinery service*, Seventh edition, 1989 (American Petroleum Institute, Washington DC).

API 674: *Positive displacement pumps, reciprocating*, 1987 (American Petroleum Institute, Washington DC).

API 675: *Positive displacement pumps, controlled volume*, 1980 (American Petroleum Institute, Washington DC).

API 676: *Positive displacement pumps, rotary*, 1980 (American Petroleum Institute, Washington DC).

GRIST, E., 'The volumetric performance of cavitating centrifugal pumps, Parts 1 and 2', *Proc. Instn mech. Engrs, Part A*, 1986, **200**, 159–167; 168–172.

ISO 5199–1986 *British Standard classification for centrifugal pumps, Class 2 (BS 6836–1987)*, 1986 (International Standards Organization).

ISO 2859–1975 *End suction pumps (rating 16 bar) – Design nominal Duty Point and dimensions*, 1975 (International Standards Organization).

KARASSIK, I., *Pump clinic*, 1981 (Marcel Dekker, New York).

NEUMANN, B., *The interaction between geometry and performance of a centrifugal pump*, 1991, (Mechanical Engineering Publications, London).

Pumping manual, Eighth edition, 1989 (Elsevier, London, Amsterdam, New York).

TILLNER, W., *et al.*, *Vermeidung von Kavitationsschaden*, 1990 (Expert Verlag, Ehningen bei Boblingen); published in English as *The avoidance of cavitation damage* (Edited by R.K. Turton), 1993 (Mechanical Engineering Publications, London).

TURTON, R.K., *Principles of rotodynamic pump design*, 1993 (Cambridge University Press).

YOUNG, F.R., *Cavitation*, 1989 (McGraw-Hill, New York).

INDEX